ADVANCES IN ELECTRON TRANSFER CHEMISTRY

Volume 6 • 1999

ADVANCES IN ELECTRON TRANSFER CHEMISTRY

Editor: PATRICK S. MARIANO
Department of Chemistry
University of New Mexico

VOLUME 6 • 1999

JAI PRESS INC.
Stamford, Connecticut

100 Prospect Street
Stamford, Connecticut 06904-0811

ISBN: 0-7623-0213-5
ISSN: 1061-8937

Transferred to digital printing 2006

CONTENTS

LIST OF CONTRIBUTORS

Kimio Akiyama — Institute for Chemical Reaction Science
Tohoku University
Sendai, Japan

Toshio Fuchigami — Department of Electrochemistry
Tokyo Institute of Technology
Yokohama, Japan

Hiroshi Ikeda — Department of Chemistry
Graduate School of Science
Tohoku University
Sendai, Japan

Hijime Maeda — Department of Applied Chemistry
College of Engineering
Osaka Prefecture University
Osaka, Japan

Tsutomu Miyashi — Department of Chemistry
Graduate School of Science
Tohoku University
Sendai, Japan

Kazuhiko Mizuno — Department of Applied Chemistry
College of Engineering
Osaka Prefecture University
Osaka, Japan

Akira Sugimoto — Department of Applied Chemistry
College of Engineering
Osaka Prefecture University
Osaka, Japan

Yasutake Takahashi — Chemistry Department of Materials
Faculty of Engineering
Mie University
Mie, Japan

Toshiyuki Tamai — Osaka Municipal Technical
Research Institute
Osaka, Japan

PREFACE

The consideration of reaction mechanisms involving the movement of single electrons is now becoming quite common in the fields of chemistry and biochemistry. Studies conducted in recent years have uncovered a large number of chemical and enzymatic processes that proceed via single electron transfer pathways. Still numerous investigations are underway probing the operation of electron transfer reactions in organic, organometallic, biochemical, and excited state systems. In addition, theoretical and experimental studies are being conducted to gain information about the factors that govern the rates of single electron transfer. It is clear that electron transfer chemistry is now one of the most active areas of chemical study.

The series, *Advances in Electron Transfer Chemistry*, has been designed to allow scientists who are developing new knowledge in this rapidly expanding area to describe their most recent research findings. Each contribution is in a minireview format focusing on the individual author's own work as well as the studies of others that address related problems. Hopefully, *Advances in Electron Transfer Chemistry* will serve as a useful series for those interested in learning about current breakthroughs in this rapidly expanding area of chemical research.

Patrick S. Mariano
Series Editor

PHOTOINDUCED ELECTRON TRANSFER REACTIONS OF CYCLOPROPANE DERIVATIVES

Tsutomu Miyashi, Hiroshi Ikeda,
Yasutake Takahashi, and Kimio Akiyama

Advances in Electron Transfer Chemistry
Volume 6, pages 1–39.

ISBN: 0-7623-0213-5

1. INTRODUCTION

Trimethylene biradicals are key intermediates in thermal and photochemical stereomutation and structural isomerization reactions of cyclopropane derivatives. As a result, a large number of thermal and photochemical reactions of cyclopropane derivatives have been investigated in an effort to understand the fundamental nature of trimethylene biradicals. Because of their low ionization potentials, highly strained cyclopropane derivatives also serve as good targets for the study of single electron transfer (SET)-promoted reactions. In fact, a variety of compounds containing cyclopropane ring systems in various environments have been investigated as substrates for photoinduced electron transfer reactions including SET-photosensitized processes and photoreactions initiated by irradiation of electron donor–acceptor (EDA) complexes. In addition, nonphotoinduced electron transfer reactions of substances in this family, initiated by γ-irradiation in a rigid matrix or chemical oxidation, have been the subject of experimental studies.

Because of the ionic nature and variable structures of cyclopropane cation radicals, the reactivity of these intermediates is often more varied than that of cyclopropane derived transients produced by pyrolysis or direct photolysis. However, in many cases the observed reaction pathways and reaction mechanisms depend on the type of electron transfer conditions employed. Thus, elucidation of a mechanistic connection between reactant structure and reactivity is a matter of concern in electron transfer reactions of cyclopropane derivatives. As the electron transfer chemistry of cyclopropanes has been reviewed recently by Roth,[1] the current discussion of reactivity profiles and structures of cyclopropane cation radicals will focus mainly on our own studies of photoinduced electron transfer photoreactions of spiropentanes and methylenecyclopropanes and a comparison of the results with those obtained from studies of direct photochemical and thermal reactions of relevant cyclopropane derivatives.

2. ELECTRON TRANSFER PHOTOREACTIONS OF CYCLOPROPANE AND SPIROPENTANE DERIVATIVES

2.1. Cyclopropanes

Two types of cyclopropane cation radicals have been suggested on the basis of both theoretical[2] and spectroscopic studies.[3,4] The first is the $^2A_1$ cation radical (**1**$^{\bullet+}$) having a one-electron σ bond. This structure is assigned by analysis of the ESR spectrum of γ-irradiated cyclopropane in a rigid matrix.[3a] The cyclopropane cation radical (**1**$^{\bullet+}$) is regarded as the immediate precursor of the second possible structure, a ring opened trimethylene cation radical (**3**$^{\bullet+}$).[3b–d] However, *ab initio* calculations do not support this proposal.[2a–c] Nevertheless, the majority of electron-transfer-induced reactions of cyclopropanes can be explained by invocation of the intermediacy of either $^2A_1$-type or trimethylene-type cation radical intermediates. Interestingly, there is no theoretical support[2] for the ring-closed $^2B_2$ cation radical (**2**$^{\bullet+}$) having the three-electron two-σ-bond structure, even though the existence of this form of the cyclopropane cation radical is suggested by the results of a chemically induced dynamic nuclear polarization (CIDNP) study by Roth.[4]

The $^2A_1$-type cation radical intermediates are frequently proposed for the nucleophilic substitution reactions of aryl-substituted cyclopropane cation radicals.[5] Rao and Hixson were the first to report the unusual regioselective nucleophilic substitution of **4**$^{\bullet+}$.[5a] The more highly substituted C-2 of **4**$^{\bullet+}$ is regioselectively attacked by methanol and reduction of the resulting radical **5**$^{\bullet}$ followed by protonation gives

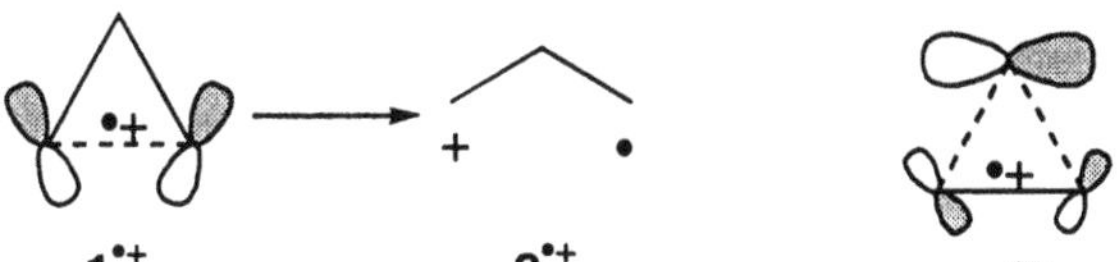

1$^{\bullet+}$ **3**$^{\bullet+}$ **2**$^{\bullet+}$

Scheme 1.

6. Recent systematic studies of Dinnocenzo confirm that steric effects[5b] in the three-electron S_N2 reaction[5b,c] are very small and that nucleophilic addition of methanol to (*S*)-$\mathbf{7}^{\bullet+}$ proceeds with complete inversion of configuration, giving rise to (*R*)-**8** (Scheme 2).

The anion radicals of the SET sensitizers, *N*-methylphthalimide,[6] chloranil (CA),[7a–c] naphthoquinone,[7c] 2,3-dichloronaphthoquinone,[7c,d] and 9,10-dicyanoanthracene (DCA)[8] undergo similar nucleophilic additions to intermediate cyclopropane cation radicals in sensitized photoreactions of arylcyclopropanes. For instance, the CA-sensitized photoreaction of **9** in the presence of methanol gives the CA/methanol adduct **12**.[7a,c] Presumably, nucleophilic substitution of $\mathbf{9}^{\bullet+}$ by methanol initially takes place to form $\mathbf{11}^{\bullet}$, to which $CAH^{\bullet}$ adds to form **12**. The steady-state photolysis of a solution of **9**, DCA, and $Cu(BF_4)_2$[9] in acetonitrile containing methanol or *t*-butanol gives **13** and **14**, respectively.[7a] Similar photoreaction of **9** in acetonitrile containing methanol and *t*-butanol gives **15** and **16**. Apparently, Cu^{2+} oxidizes not only $DCA^{\bullet-}$ but also radical $\mathbf{11}^{\bullet}$.[9] Under these sensitized conditions, the rate constants[10] for the nucleophilic substitution, k_{MeOH} and $k_{t\text{-}BuOH}$, are determined to be 1.2×10^7 and $5.3 \times 10^6\ dm^3mol^{-1}s^{-1}$, respectively.[7a]

In contrast, photoexcitation of the EDA complex of **9** and CA does not lead to product formation.[7b] Rapid back electron transfer from

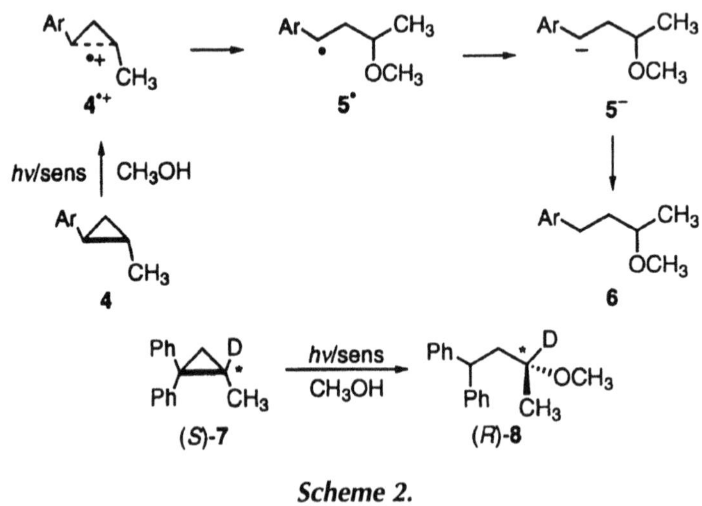

Scheme 2.

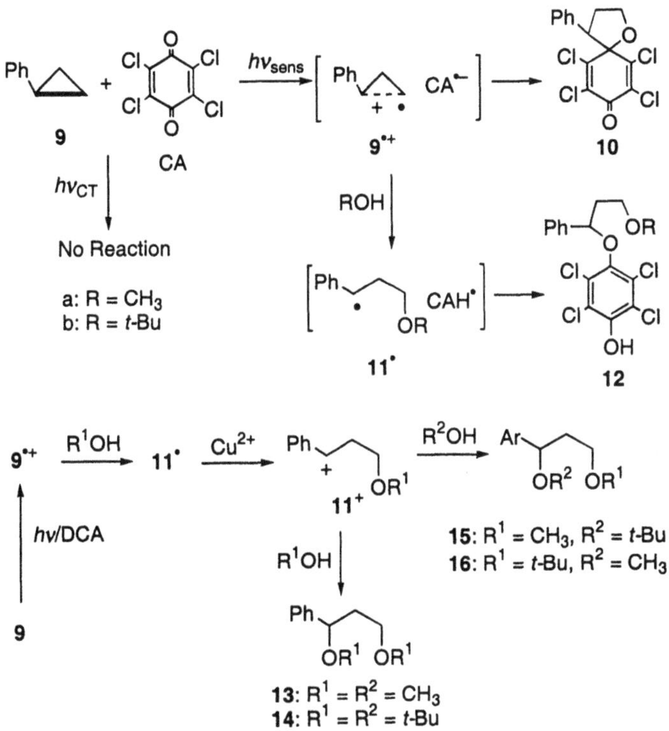

Scheme 3.

$CA^{\bullet -}$ to $9^{\bullet +}$ is apparently responsible for this phenomenon. However, photoexcitation of the EDA complex of 1,1-diphenylcyclopropane (**17**) and CA does give rise to an SET reaction to generate **19**. Ion pair

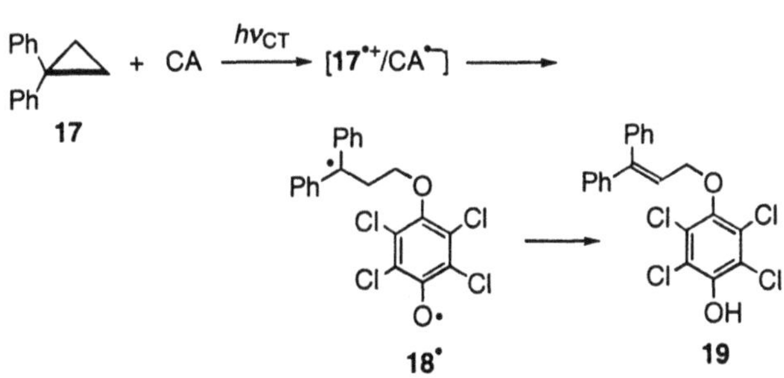

Scheme 4.

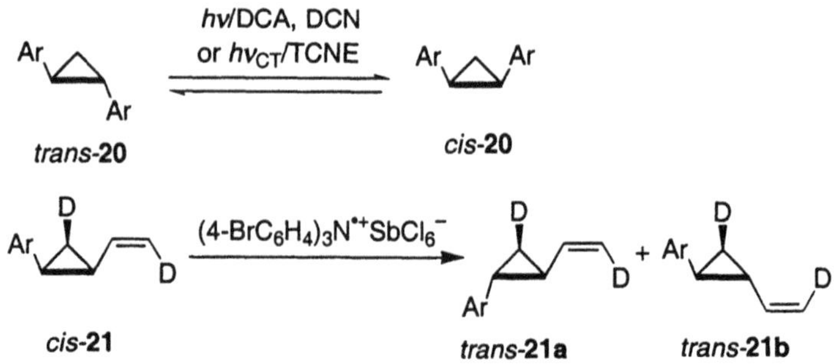

Scheme 5.

coupling[6] within the ion radical pair [$17^{\bullet+}$/$CA^{\bullet-}$] followed by a H-shift is likely the mechanistic route followed in this process. Picosecond laser flash photolysis has been employed to confirm the intermediacy of the transient ion radical pair [$17^{\bullet+}$/$CA^{\bullet-}$] in this reaction.[7b]

Stereomutation of 1,2-diarylcyclopropane (**20**) takes place under DCA,[11] 1,4-dicyanonaphthalene (DCN),[12] or 3,3′,4,4′-benzophenonetetracarboxylic dianhydride (BTDA) SET-sensitized[13] conditions as well as when the EDA complex of **20** and tetracyanoethylene (TCNE) is directly irradiated.[14] Interestingly, the aminium salt catalyzed reaction of *cis*-**21** also involves stereomutation with double epimerization, affording *trans*-**21a** and *trans*-**21b**.[15]

Stereomutation of **22** can be formally explained by invoking the intermediacy of the trimethylene cation radical $\mathbf{23}^{\bullet+}$. Nevertheless, two intriguing mechanistic alternatives for this process have been proposed.[12,13] Based on CIDNP and energetic considerations, Wong and Arnold proposed that the stereomutation of *trans*-**22** proceeds through the triplet trimethylene biradical $\mathbf{23}^{\bullet}$ which is formed by back electron transfer from $DCN^{\bullet-}$ to $\mathbf{23}^{\bullet+}$ within the triplet ion radical pair $^3[\mathbf{23}^{\bullet+}/DCN^{\bullet-}]$.[12] On the other hand, Dinnocenzo and his co-workers recently proposed a dissociative return electron transfer mechanism.[13] In their route, ring cleavage of *trans*-$\mathbf{22}^{\bullet+}$ and back electron transfer from $BTDA^{\bullet-}$ to *trans*-$\mathbf{22}^{\bullet+}$ occurs within the triplet ion radical pair to form simultaneously the triplet $\mathbf{23}^{\bullet}$, through which the stereomutation of *trans*-**22** takes place to yield *trans*-**22** and *cis*-**22**. Although it would be interesting to know which mechanism is really operative for aryl-

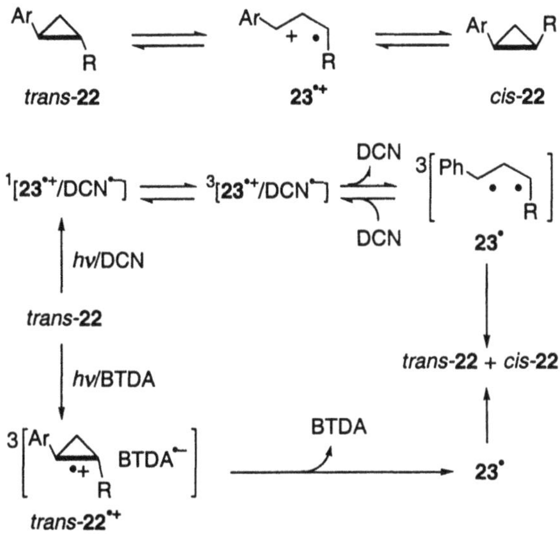

Scheme 6.

cyclopropane cation radical stereomutation, it is noteworthy that a common feature of the latter two mechanisms includes the back electron transfer process to form a trimethylene biradical intermediate.

Structural isomerizations of arylcyclopropane cation radicals to propene derivatives are also frequently observed.[16] Trimethylene cation radicals are the most probable intermediates in these processes. However, in some cases the nature of the mode used to promote the SET event is crucial in determining the type of rearrangement sequence that is followed. For instance, DCN-sensitized photoreaction of **24** gives **25**, whereas photoexcitation of the EDA complex of **24** and TCNE leads to production of **26**.[16a] Oxygenation reactions[11,16b,c,17] to form dioxolanes are also typical reactions of arylcyclopropane cation radicals formed under aerobic conditions. Of particular interest are the DCA-sensitized oxygenation reactions of *cis*-**27** and *trans*-**27**, both of which give *cis*-**28** as a major product.[11,17a] Addition of biphenyl as a cosensitizer[17a] or magnesium perchlorate[11] is known to accelerate these processes. Trimethylene cation radicals are plausible intermedi-

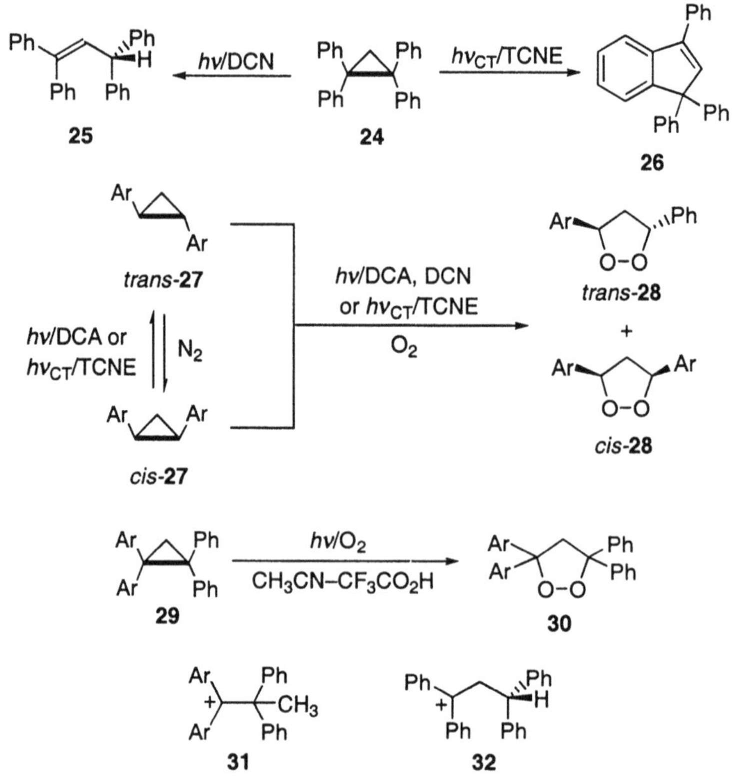

Scheme 7.

ates, but different types of oxygen species have been proposed as the oxidant in these systems.[16b–d,17]

Photoexcitation of the EDA complex of *cis*-**27** or *trans*-**27** and TCNE under an oxygen atmosphere also lead to formation of *cis*-**28** as the major product.[17b] In these cases, the trimethylene cation radicals formed from $\mathbf{27}^{\cdot+}$ are captured by molecular oxygen and the ensuing stepwise cyclization then yields **28**. Interestingly, oxygenation to form **30** occurs when an acetonitrile solution of **29** is irradiated ($\lambda > 390$ nm) under an oxygen atmosphere in the presence of trifluoroacetic acid.[17c] Presumably, the carbenium ion **31** or **32** is initially formed from **29** and the photoexcited **31** or **32** then sensitizes oxygenation.

No chemical reactions have been observed which require the intermediacy of the $^2B_2$-type cation radical. However, CIDNP experiments with **33** provide evidence for the existence of the unique structure of **33**$^{\bullet+}$ in which two cyclopropane bonds are simultaneously weakened.[4] It is noteworthy that spiroannelation with the fluorene ring causes a remarkable change in the distribution of the charge and odd electron densities in the ring of cyclopropane cation radicals.

In connection with the spiroannelation phenomenon, the final example in this section focuses on the reactivities of bicyclo[6.1.0]nonatriene cation radicals. Spiroannelation at C-9 with a fluorene ring dramatically changes the rearrangement pathway followed in their SET reactions.[18] CA-sensitized photoreaction of **34a** gives 1,3,5,7-cyclononatetraene (**38**) whereas the DCA-sensitized photoreaction of the diphenyl derivative **34b** affords the cycloheptatriene derivative **39** as the sole product. Remarkably, the DCA-sensitized photoreaction of **34c** gives the barbaralane derivative **40** as a major product.

A similar rearrangement occurs when the EDA complex of **34c** and TCNE is irradiated.[18a] This EDA complex in dichloromethane has charge-transfer (CT) absorption maxima at 402 and 568 nm which can be ascribed to SET interactions between the fluorene ring and TCNE. Pertinent to this is the fact that the EDA complex of fluorene and TCNE exhibits CT absorption maxima at 420 and 560 nm in dichloromethane. Photoexcitation ($\lambda > 400$ nm) of this EDA complex gives rise to generation of **42** which is the secondary adduct of **40** and TCNE. As the thermal reaction of **34c** and TCNE affords **43**, the striking difference between photochemical and thermal reactions supports the operation of an electron transfer mechanism for the rearrangement of **34c** to **40**. The difference in the reactivity of **34a**, **34b**, and **34c** can be reasonably explained by differences in charge and odd electron distributions in the cation radical intermediates. CIDNP experiments reveal that the fluorene moiety is the primary electron donor site and that the cation radical is localized predominantly on the fluorene ring of **37**$^{\bullet+}$. The cation radical distributions in **35**$^{\bullet+}$ and **36**$^{\bullet+}$ are shown in Scheme 8. Rearrangement to give **40** can be explained by intervention of the highly stabilized homotropylium cation radical. The C-8–C-9 bond of **37**$^{\bullet+}$ is initially cleaved to form *anti*-**41**$^{\bullet+}$ and the successive bond formation in *syn*-**41**$^{\bullet+}$ gives **40**.

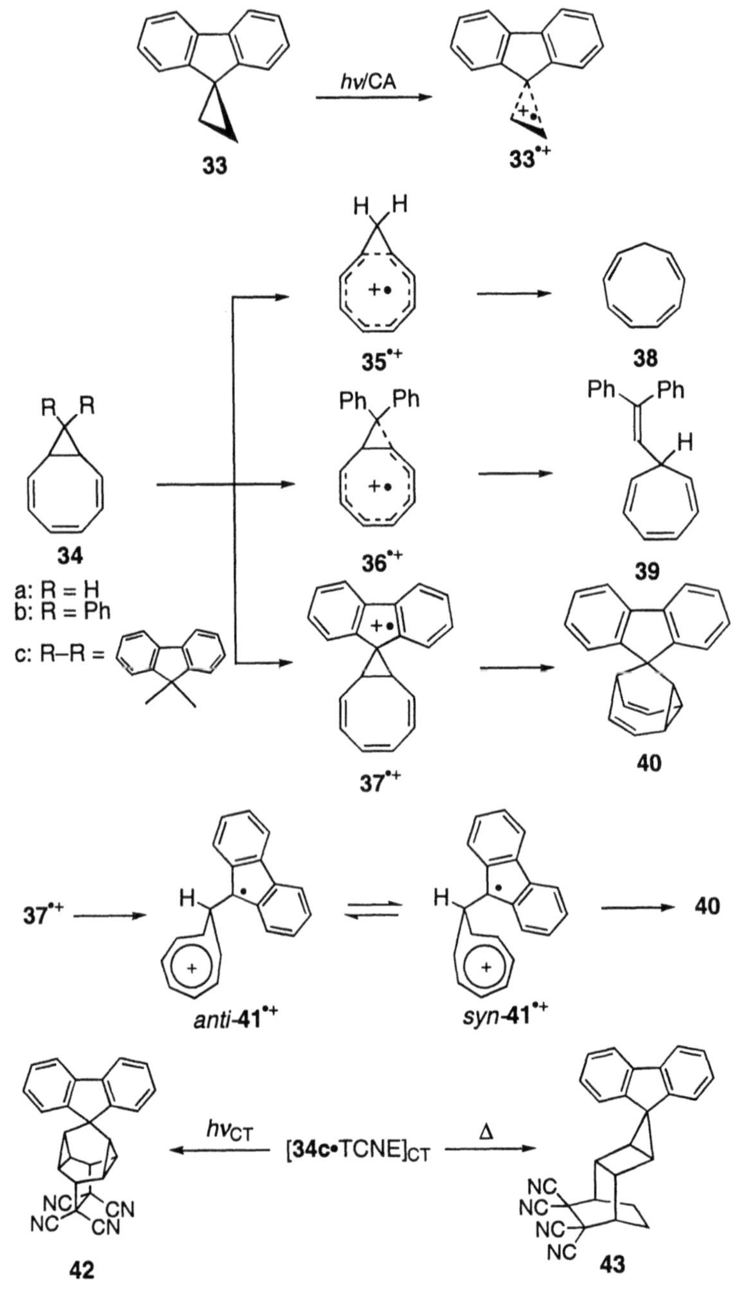
hν/CA
33
33•+
H
H
+•
35•+
38
R
R
34
a: R = H
b: R = Ph
c: R–R =
Ph
Ph
36•+
39
37•+
40
anti-41•+
syn-41•+
hνCT
[34c•TCNE]CT
Δ
NC
CN
42
43

Scheme 8.

2.2. Spiropentanes

On pyrolysis, spiropentane (**44**) undergoes a spiropentane–methylenecyclobutane rearrangement[19,20] to form **45**.[21] This process involves two successive bond cleavage reactions. Initial peripheral bond cleavage forms the cyclopropylbiscarbinyl biradical **46**• which then rearranges to the 1,4-biradical **47**• followed by closure. Biradical **47**• is also known to be a key intermediate in the degenerate methylenecyclobutane rearrangement of **45**.[22] The rearrangement of **44** to **45** is alternatively explained by the initial radial bond cleavage to form **48**•. The successive cyclopropyl–allyl rearrangement of **48**• to **47**• followed by closure gives **45**. No experimental evidence has been provided for the radial bond fission of **44** on pyrolysis, but **44** and **45** are known to be formed from the vibrationally excited state of **48**• generated on pyrolysis of **49** in the gas phase.[23] The cation radical spiropentane–methylenecyclobutane rearrangement of **44** was reported by Shida and his co-workers.[24] Cation radical **44**•+ generated by γ-irradiation in a Freon matrix rearranges to form **45**•+, but the possible cation radical intermediates corresponding to **46**•, **47**•, or **48**• could not be detected by electron paramagnetic resonance (EPR) spectroscopic methods.

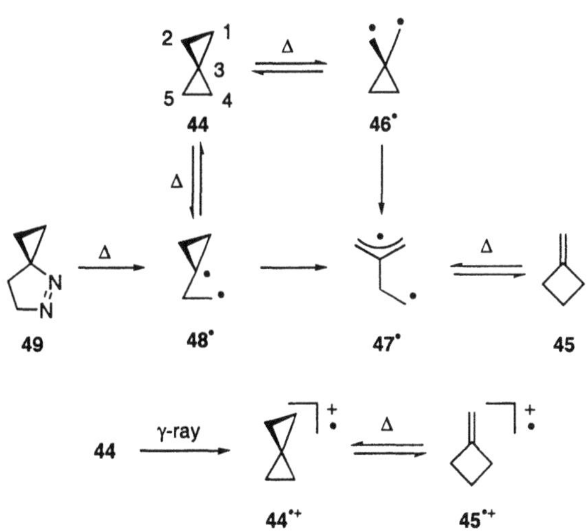

Scheme 9.

Direct and acetone triplet-sensitized photolyses of 2,2-diarylspiropentanes (**50a–c**) also promote the photochemical spiropentane–methylenecyclobutane rearrangement.[25] For instance, irradiation of **50a–c** with 254-nm light gives methylenecyclobutanes (**51a–c**, **52a–c**, and **53a–c**) together with 1,1-diarylallenes (**54a–c**) in the yields shown in Table 1. The formation of methylenecyclobutanes is formally accounted for by the mechanism shown in Scheme 10. Initial peripheral bond cleavage leads to formation of **51** and **52** whereas a radial bond cleavage route yields **53**. The formation of **54** from **50** resembles the reaction[23] of the vibrationally excited state of **44** which generates ethylene and allene via the vibrationally excited state of **45**. However, the methylenecyclobutanes, **51**, **52**, and **53**, are not the immediate precursor of **54**. A plausible intermediate in this reaction pathway is the cyclopropylcarbene **59**, formed by the ethylene extrusion process. Nevertheless, the reactivity of the excited states of **50** formed by direct irradiation resemble those of the vibrationally excited ground states of **44** and **45**. The acetone-sensitized photoreaction of **50** occurring through its triplet excited state proceeds differently, affording only **51** and **52** in the yields shown in Table 1.

The electron transfer spiropentane–methylenecyclobutane rearrangement of **50** to **51** and **52** occurs under a variety of photochemical conditions, but unlike in the direct photolytic reaction, neither **53** nor **54** is formed.[25b,26,27] Spiropentanes (**50a–c**) are good electron donors and form colored EDA complexes with TCNE.[25b] The EDA complex of the more potent electron donor, **50a** ($E^{ox} = +1.17$ V vs. SCE), and TCNE exhibits CT absorption maxima at 398 and 598 nm in dichloromethane. The long-wavelength band of the less electron-rich do-

Table 1. Direct[a] and Acetone-Sensitized[b] Photoreactions of **50**

	Direct, Yields (%)					*Sens., Yields (%)*		
Sub	**51**	**52**	**53**	**54**	*Rec*	**51**	**52**	*Rec*
50a	12	20	2	12	37	21	45	9
50b	12	11	6	35	31	27	40	22
50c	10	9	10	38	30	21	27	38

Notes: [a]Solutions of **50** (0.12 mmol) in acetonitrile (6 mL) were photolyzed (254 nm) for 3 h.
[b]Solutions of **50** (0.2 mmol) in acetone (6 mL) were photolyzed (300 nm) for 5 h.

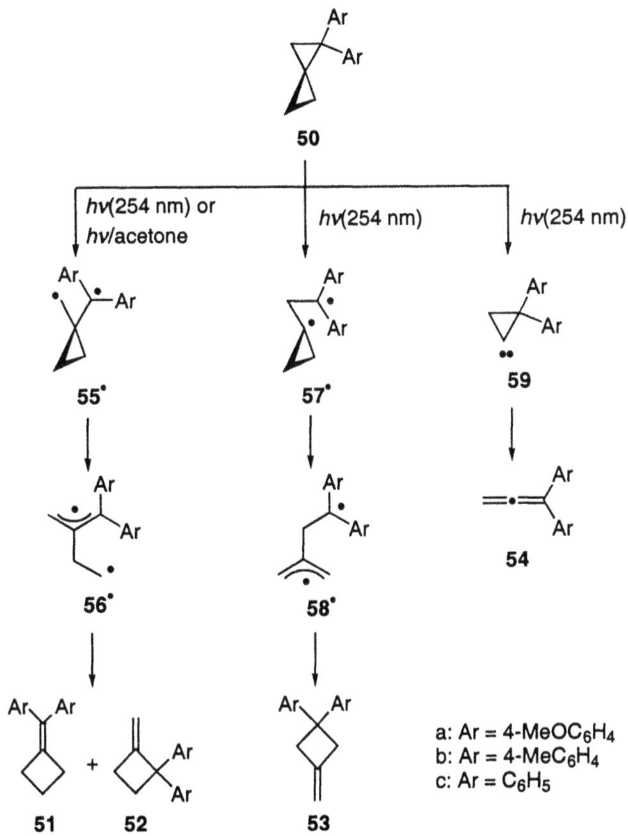

Scheme 10.

nors, **50b** (E^{ox} = +1.42 V) and **50c** (E^{ox} = +1.67 V), are significantly blueshifted (419 and 540 nm for **50b** and 406 and 486 nm for **50c**) as expected from an increase in their oxidation potentials.

Photoexcitation (λ > 390 nm) of the EDA complex between **50a** and TCNE affords **51a** and **52a**, and the TCNE adducts **60a** and **61a** together with **62a**. Photoreaction of the more weakly electron-donating **50c** gives predominantly **51c** (Table 2). Interestingly, the efficiency of formation of the TCNE adducts appears to depend on the electron-donating ability of the spirocyclopentane **50** while the **51**:**52** product ratio seems to depend on the polarity of the solvent. Within the initially formed ion radical pair [**50a**$^{•+}$/TCNE$^{•-}$], peripheral bond cleavage to

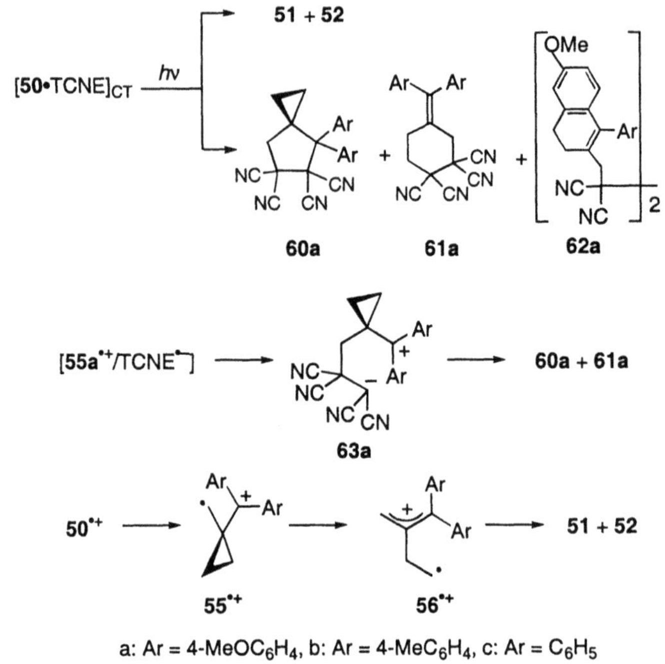

Scheme 11.

form the ion radical pair [$55a^{•+}$/$TCNE^{•-}$] followed by rapid capture of $55a^{•+}$ by $TCNE^{•-}$ occurs to generate **63a**, which is stabilized by the electron-donating methoxy substituent. Closure of **63a** gives **60a** while rearrangement of **63a** followed by cyclization forms **61a**. On the other hand, the formation of **51** and **52** can be formally explained

Table 2. Photoreactions of EDA Complexes of **50** and TCNE[a]

			Conversion Yields (%)				
Sub	*Solvent*	*Time (h)*	**51**	*52*	**60**	**61**	**62**
50a	CH_2Cl_2	6	5	3	23	11	45
	CH_3CN	6	37	0	8	4	21
50c	CH_2Cl_2	20	68	11	0	0	0
	CH_3CN	20	95	0	0	0	0

Note: [a]4-mL dichloromethane solutions of **50** and TCNE (0.2 mmol) were irradiated (λ > 390 nm).

by closure of **56**$^{\bullet+}$ in two different ways, but this single mechanism cannot explain the solvent polarity dependence of the ratio of **51** and **52**.

Similar solvent dependencies are observed in the CA and anthraquinone (AQ) SET-sensitized photoreactions of **50**.[25b,26,27] The CA-sensitized irradiation of **50** leads to production of **51** and **52** along with the CA adduct **64**, a secondary photoproduct of **51**. These results together with those of the AQ-sensitized photoreaction are summarized in Table 3. Apparently, an increase in the solvent polarity leads to a large increase in the relative ratio of **51** to **52** in the CA-sensitized process and only a moderate increase under the AQ-sensitized conditions. Of particular interest is the fact that oxygenation reaction products, **65** and **66**, are produced only under those SET-sensitized conditions in which **52** is formed. For instance, irradiation of a dichloromethane solution of AQ and **50a**, pressurized with oxygen (20 kg/cm^2), gives **65a** (7%) and **66a** (8%) in poor yields along with **52a** (27%) while CA-sensitized photoreaction of **50a** in acetonitrile does not yield any oxygenation product. This observation suggests that **52**, **65**, and **66** most likely originate from a common precursor.

An overview of the results presented above leads to the proposal that the solvent dependence is most likely a consequence of the initial formation of two different types of spirocyclopentane cation radical intermediates. The degree of interaction between the substrate cation radical and sensitizer anion radical should vary in a continuous fashion

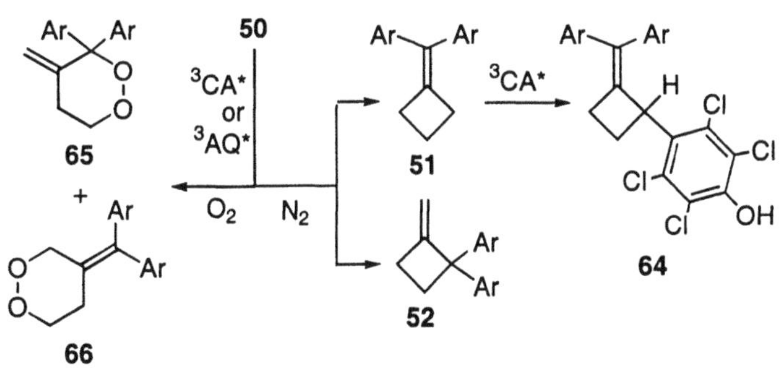

a: Ar = 4-MeOC$_6$H$_4$, b: Ar = 4-MeC$_6$H$_4$, c: Ar = C$_6$H$_5$

Scheme 12.

Table 3. Photoreactions of **50** under AQ-, CA-Sensitized Conditions[a]

		Yields and Conversions (%)						
		AQ Sensitization			*CA Sensitization*			
Sub	*Solvent*	**51**	**52**	*Con*	**51**	**52**	**64**	*Con*
50a	CH_3CN	33	26	79	3	2	45	100
	CH_2Cl_2	17	41	68	6	< 2	12	100
	C_6H_6	12	43	86	0	40	0	100
50b	CH_3CN	19	29	61	8	< 2	72	100
	CH_2Cl_2	13	31	59	18	10	44	99
	C_6H_6	0	4	13	0	32	< 2	100
50c	CH_3CN	13	29	59	18	4	49	100
	CH_2Cl_2	< 2	18	28	6	13	47	94
	C_6H_6	< 2	11	18	5	31	5	97

Note: [a]A 5-mL solution of **50** (0.08 mmol) and AQ (0.01 mmol) or CA (0.08 mmol) was irradiated for 30 min (AQ) or 60 min (CA).

with solvent polarity and, thus, the distribution of positive charge and unpaired electron density in the substrate cation radical should be altered by the polarity of the solvent.

In more polar solvents, $\mathbf{50}^{\bullet+}$ exists either as a free cation radical or as the cation in a solvent-separated ion radical pair. The positive charge and unpaired electron of $\mathbf{50}^{\bullet+}$ are delocalized over the elongated peripheral bond. This electronic distribution drives C-4 migration to C-2 to form **51** concertedly. The process, facilitated by an increase in the solvent polarity, preferentially operates in the CA-sensitized pho-

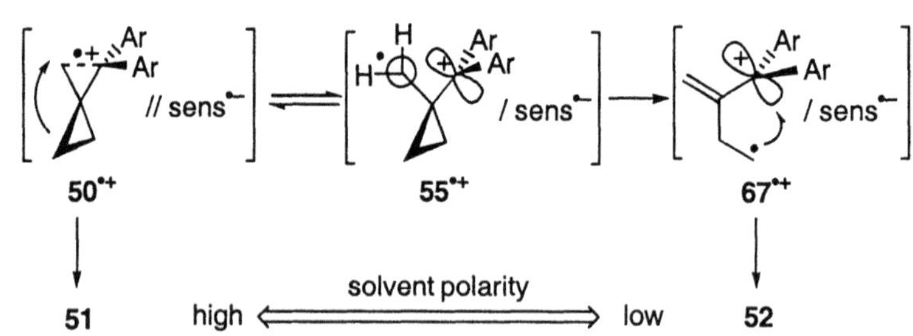

Scheme 13.

toreaction of **50a** and CT-irradiation of the EDA complex of **50c** and TCNE in polar acetonitrile. Molecular oxygen does not capture **50a**$^{\bullet+}$ even in acetonitrile. Rather, in less polar solvents, **50**$^{\bullet+}$, as part of a contact ion radical pair, rearranges to **67**$^{\bullet+}$ through **55**$^{\bullet+}$ in which the positive charge and unpaired electron are completely separated. This sequential rearrangement process strengthens the ion radical pair interaction and leads to electrostatic stabilization of the contact ion radical pairs [**55**$^{\bullet+}$/sens$^{\bullet-}$] and [**67**$^{\bullet+}$/sens$^{\bullet-}$].

Like in the diaryl-substituted trimethylenemethane cation radical (Section 3), the *p*-orbital at C-1 in **67**$^{\bullet+}$ is probably perpendicular to the C-2–C-3 π-orbital because of the bulky aryl groups. The ring closure at C-1 and C-4 thus selectively forms **52**. This process operates preferentially in less polar solvents, especially in the nonpolar benzene (Tables 2 and 3). The formation of **65** and **66** in low yields under aerobic conditions can be accounted for by molecular oxygen capture of **67**$^{\bullet+}$ which reluctantly diffuses from the contact ion radical pair [**67**$^{\bullet+}$/sens$^{\bullet-}$].

The use of stereochemical tests provided in the reactions of *trans*-**68**, *anti,cis*-**68**, and *syn,cis*-**68** leads to support for the operation of diverse rearrangement mechanisms.[26,27] For example, the CA SET-sensitized photoreactions of *trans*-**68** and *syn,cis*-**68** afford *trans*-**69** and *cis*-**69**, respectively. However, the 2,2-diphenyl-1-methylenecyclobutane products are generated as stereoisomeric mixtures (*trans*-**72** and *cis*-**72**) (Scheme 14). Because the methyl group of *anti,cis*-**68** is oriented toward the C-4 migration site, *anti,cis*-**68** does not rearrange to yield *cis*-**69**, but rather gives different products. Thus, the solvent polarity-dependent diverse reactivity of **50**$^{\bullet+}$ is a characteristic feature of the photoinduced electron transfer reactions of **50**.

The three methylenecyclobutanes, **51**, **52**, and **53**, do not interconvert under direct irradiation or SET-photosensitized conditions. However, 3,3-diaryl-4,4-dideuterio-1-methylenecyclobutane (d_2-**53**) does undergo electron-transfer photoinduced degenerate methylene-cyclobutane rearrangement when either DCN, DCA, or 2,6,9,10-tetracyanoanthracene is used as sensitizer. These processes involve the allylically stabilized 1,4-cation radical intermediates d_2-**58**$^{\bullet+}$ and d_2-**58**$'^{\bullet+}$.[28] Here, the 1,4-cation radical **58**$^{\bullet+}$ is efficiently captured by molecular oxygen, giving rise to **73** as shown in Scheme 15.

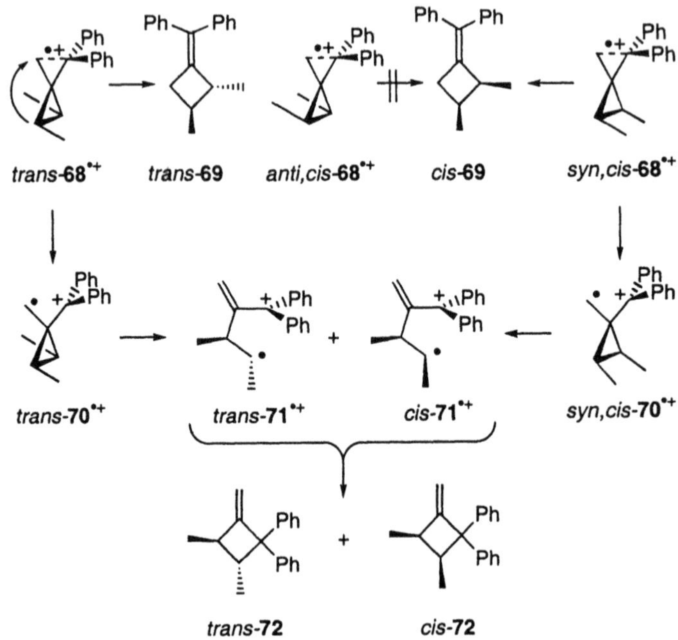

Scheme 14.

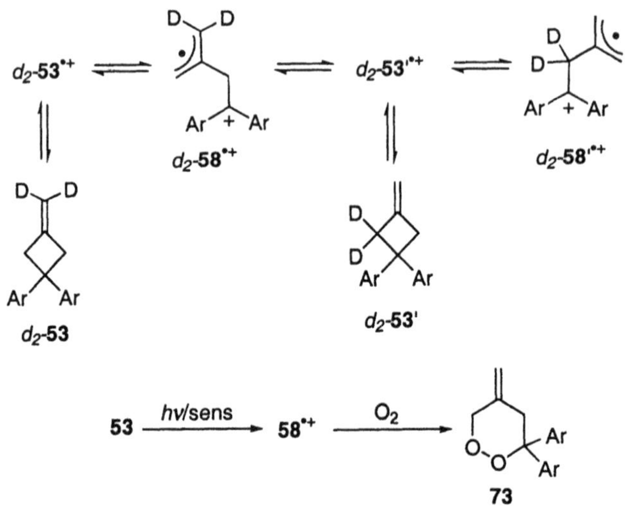

a: Ar = 4-MeOC$_6$H$_4$, b: Ar = 4-MeC$_6$H$_4$, c: Ar = C$_6$H$_5$

Scheme 15.

3. ELECTRON TRANSFER PHOTOREACTIONS OF METHYLENECYCLOPROPANE DERIVATIVES

3.1. The Electron Transfer Photoinduced Degenerate Methylenecyclopropane Rearrangement

Since the time when the thermally induced methylenecyclopropane rearrangement (**A** → **B**) of Feist's esters was first observed by Ullman in 1959,[29] rearrangements of a large number of methylenecyclopropane derivatives have been subjected to kinetic, stereochemical, and theoretical studies. The main objectives of these efforts were to understand the role and nature of the trimethylenemethane biradical intermediate ($\mathbf{C}^{\bullet}$) in the rearrangement process.[30] Considerable attention has focused on the theoretical[31] and experimental[32] elucidation of the relationship among structure, spin state, and reactivity of this simple non-Kekulé molecule as well as on its applications as a synthetic[33] and practical[34] intermediate.

In contrast to the thermal process, only a limited number of studies of the photochemical methylenecyclopropane rearrangement have been reported. For instance, photolysis[35,36] of methylenecyclopropane derivatives is known to induce this rearrangement reaction. The trimethylenemethane biradical intermediate in these processes can be generated by several different procedures including photolysis of methylenecyclobutanone in a Freon matrix,[37] γ-irradiation of methylenecyclopropane in a Freon matrix,[38] and low-temperature photolysis of methylenecyclopropane in a halogen-doped xenon matrix.[39] Generation of the parent trimethylenemethane cation radical ($\mathbf{75}^{\bullet+}$) was only recently reported.[40] γ-Irradiation of **74** in a $CF_2ClCFCl_2$ matrix at 77 K forms $\mathbf{75}^{\bullet+}$. Interestingly, $\mathbf{75}^{\bullet+}$ is converted to 2-methylenecyclopropyl radical (**76**) at 120 K by a deprotonation process which is similar to that observed for the cyclopropylidenecyclopropane cation radical ($\mathbf{95}^{\bullet+}$) (see Scheme 26, page 34).

Although SET photochemistry is often a practical method to promote cation radical reactions, such a variant had not been probed until we recently discovered the SET-photoinduced degenerate methylenecyclopropane rearrangement of 2,2-diaryl-1-methylenecyclopropanes.[41] Below, we describe in detail the results of our studies of this process including the chemical and spectroscopic identification

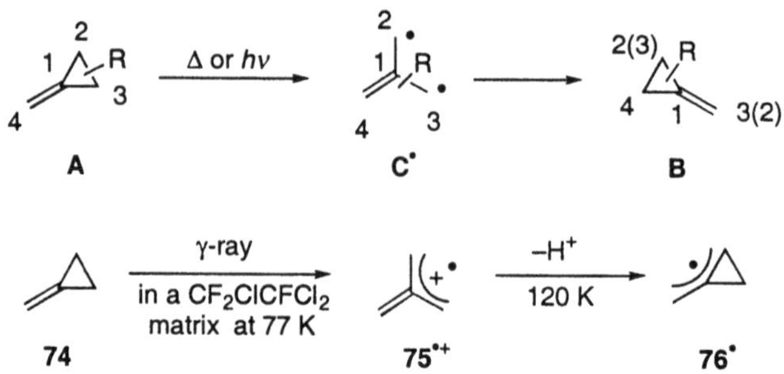

Scheme 16.

of trimethylenemethane cation radical and biradical intermediates, as well as the mechanism and energetics of the degenerate methylenecyclopropane rearrangement.

Photoexcitation of the EDA complexes of methylenecyclopropanes and TCNE is a convenient procedure to generate trimethylenemethane cation radical intermediates. 2,2-Diaryl-1-methylenecyclopropanes (**77a–c**) form colored EDA complexes when mixed with TCNE in dichloromethane.[41a] Irradiation ($\lambda > 390$ nm) of an oxygen-saturated dichloromethane solution (λ^{CT}max = 370 and 568 nm) of **77a** ($E^{ox} = +1.35$ V vs. SCE) and TCNE, leads to formation of dioxolanes **78a** and **79a** and the TCNE adducts **80a** and **81a** in respective yields of 42 and 54%.

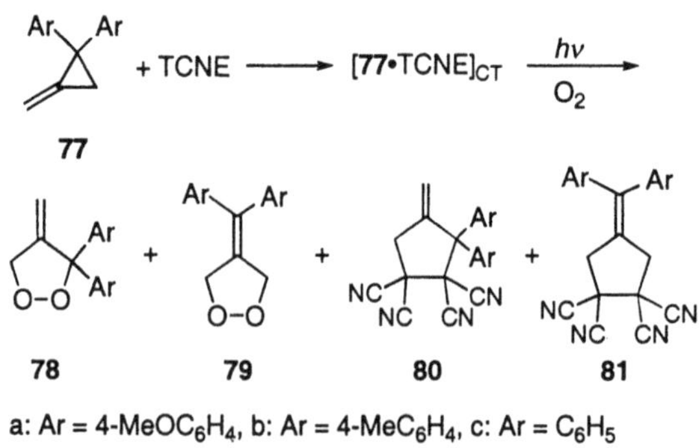

Scheme 17.

Interestingly, the relative yields of the dioxolanes and the TCNE adducts significantly depend on solvent polarity. In nonpolar benzene, the TCNE adducts are produced in an 84% yield in preference to the dioxolanes (8%), whereas in the more polar nitromethane and acetonitrile, the dioxolanes are formed exclusively. Similar results are obtained from studies of the photoreactions of the EDA complexes of **77b–c** and TCNE (λ^{CT}max = 366 nm for **77b** and 398 nm for **77c**). The findings suggest that oxygenation to form dioxolanes is an out-of-cage process whereas the formal [3+2] cycloaddition to yield TCNE adducts is an in-cage process. One intriguing finding is that the addition of 1,2,4,5-tetramethoxybenzene (TMB) as a quencher suppresses the oxygenation process completely but does not alter the [3+2] cycloaddition reaction.

On the basis of these results, the plausible mechanism shown in Scheme 18 has been formulated. Irradiation of the EDA complex

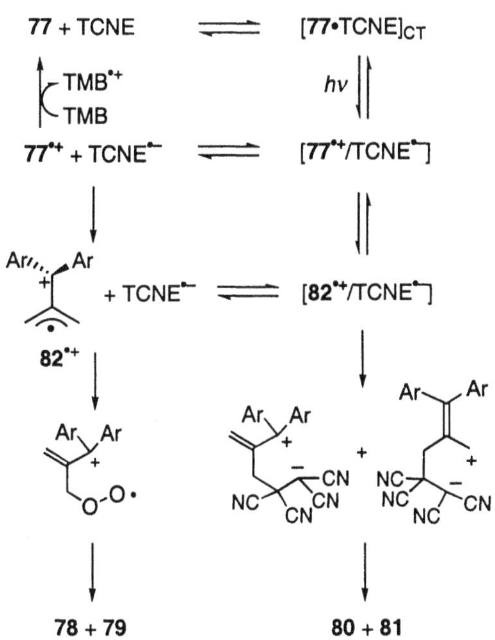

Scheme 18.

[**77**/TCNE] forms the ion radical pair [$\mathbf{77}^{\bullet+}$/TCNE$^{\bullet-}$], which subsequently collapses by either diffusion separation to form free $\mathbf{77}^{\bullet+}$ and TCNE$^{\bullet-}$ or formation of the ion radical pair [$\mathbf{82}^{\bullet+}$/TCNE$^{\bullet-}$] via C-2–C-3 bond cleavage of $\mathbf{77}^{\bullet+}$. The former and latter processes will be facilitated and suppressed, respectively, by an increase in solvent polarity. In the absence of TMB the free $\mathbf{77}^{\bullet+}$ undergoes cleavage to generate $\mathbf{82}^{\bullet+}$ which subsequently reacts with oxygen to eventually yield **78** and **79**. Free $\mathbf{77}^{\bullet+}$ can be quenched by TMB, hence suppressing the yields of oxygenation products. In contrast, [3+2] cycloaddition occurs readily within the ion radical pair [$\mathbf{82}^{\bullet+}$/TCNE$^{\bullet-}$] to form **80** and **81**.

Similar electron transfer-induced oxygenation and [3+2] cycloaddition reactions occur when the EDA complexes of **77a–c** and TCNE are stirred under oxygen in the dark at ambient temperature. This result suggests that an electron transfer mechanism is also operable in the thermal [3+2] cycloaddition of **77** and TCNE.[42]

The thermal degenerate methylenecyclopropane rearrangement of d_2-**77c** was reported by Gilbert and Butler in 1970.[43] One intriguing finding in this effort is that pyrolysis of d_2-**77c** gives an equilibrium mixture of d_2-**77c** and d_2-**77c′** in a ratio of 1.04:1.00, but it does not form d_2-**83c**, the more thermodynamically stable isomer. These results suggest the intermediacy of the bisected trimethylenemethane biradical intermediate d_2-$\mathbf{82c}^{\bullet}$ in the thermal reaction pathway. On the other hand, it is not certain whether $\mathbf{82c}^{\bullet}$ is produced on photolysis of **77c** because **83c** itself undergoes photochemical methylenecyclopropane rearrangement to give **77c**, but **77c** forms only 1,1-diphenylethylene under similar irradiation conditions.[35]

Information gathered in studies of electron transfer photoreactions of d_2-**77** demonstrates that degenerate rearrangement occurs under a variety of SET-sensitized conditions. Although **77c** efficiently quenches the luminescence of AQ, no chemical change of **77c** is observed when an acetonitrile solution of **77c** and AQ is irradiated. AQ-sensitized photoreaction of d_2-**77c** gives rise to an approximately 1:1 photostationary state mixture of d_2-**77c** and d_2-**77c′**, excluding d_2-**83c** as an intermediate in this valence bond isomerization.[41b] CA photosensitization similarly promotes the degenerate rearrangement, but neither phenanthraquinone (PQ) nor benzophenone (BP) serve as photosensitizers in this process. As electron transfer from **77c** to ^{3}CA*

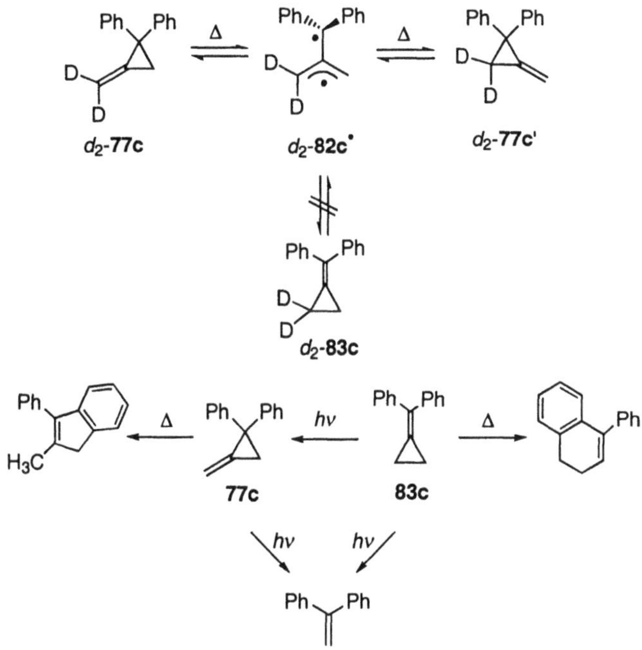

Scheme 19.

or $^3AQ^*$ is exothermic whereas that to $^3PQ^*$ or $^3BP^*$ is endothermic, the degenerate rearrangement of d_2-**77c** is reasonably accounted for by an electron-transfer mechanism.

In fact, oxygenation products **78c** and **79c** form in the AQ- and CA- (but not under the PQ- or BP-) sensitized photoreaction of **77c** in oxygen-saturated acetonitrile. In contrast, the highly electron rich anisyl derivative **77a** is oxygenated even when PQ and BP are used as SET sensitizers. The data in Table 4 show the relationships that exist between the yields of dioxolane products and the free energy changes (ΔG) associated with electron transfer from **77** to triplet excited states of the sensitizers.

2,2-Diaryl-1-methylenecyclopropanes (**77a–c**) also efficiently quench the fluorescence of singlet SET sensitizers such as DCA, 1,2,4,5-tetracyanobenzene (TCNB), and singlet cationic sensitizers such as *N*-methylquinolinium tetrafluoroborate ($NMQ^+BF_4^-$). Degenerate rearrangement and oxygenation of **77** also occur in the SET-pro-

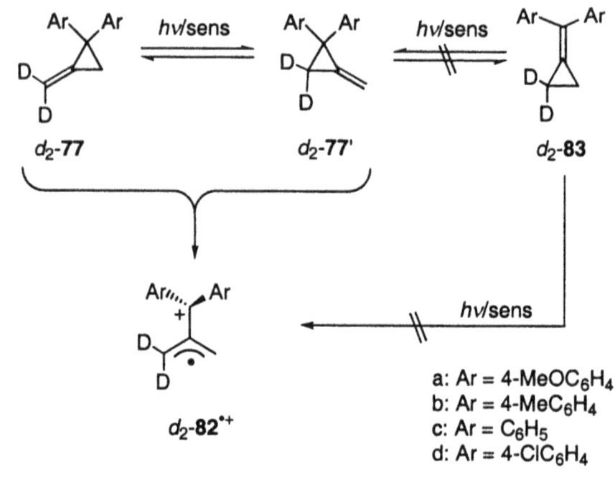

Scheme 20.

moted photoreactions using these singlet-state sensitizers.[41c] In acetonitrile, DCA-sensitized photoreactions of d_2-**77a** give a 58:42 photostationary mixture of d_2-**77a** and d_2-**77a′** under argon and dioxolane **78a** quantitatively under oxygen. While similar sensitized photoreactions in benzene or dichloromethane induce degenerate rearrangement of d_2-**77a**, in these cases the oxygenation process is extremely inefficient. TCNB-sensitized photo-degenerate rearrangement of d_2-**77a** proceeds under an argon atmosphere and oxygenation occurs quantitatively under oxygen. Like the DCA-sensitized photoreaction in dichloromethane, the degenerate rearrangement of d_2-**77a** is induced by TCNB sensitization in dichloromethane. However, oxygenation does not occur in this solvent. In contrast, under $NMQ^+BF_4^-$–toluene-cosensitized conditions in acetonitrile, degenerate rearrangement of d_2-**77a** proceeds slowly, while oxygenation rapidly takes place. The combined results suggest that degenerate rearrangement and oxygenation proceed through different intermediates in these processes.

Spectroscopic Evidence for Bisected Trimethylenemethane Cation Radical and Biradical Intermediates

Additional findings provide a clue to the detailed mechanism of the degenerate methylenecyclopropane rearrangement. The structures of cation radicals **82**$^{•+}$ and **83**$^{•+}$ were investigated using the CIDNP

Table 4. Calculated ΔG (kcal/mol) and Yields (%) of **78** and **79** in the Photosensitized Reactions of **77**

	CA Sensitization				*AQ Sensitization*				*PQ Sensitization*				*BP Sensitization*			
Sub	ΔG	**78a**	**79a**	*Con*[a]	ΔG	**78b**	**79b**	*Con*[a]	ΔG	**78c**	**79c**	*Con*[a]	ΔG	**78d**	**79d**	*Con*[a]
77a	–33	56	< 2	66	–11	58	4	73	–5.0	40	3	64	+0.2	40	< 2	53
77b	–26	62	10	85	–4.4	43	13	69	+1.9	17	< 2	32	+7.1	0	0	b
77c	–21	53	14	79	–0.2	25	8	40	+6.0	< 2	< 2	8	+11.3	0	0	b
77d	–20	51	19	81	+0.9	20	9	44	+7.2	0	0	b	+12.4	0	0	b

Notes: [a]Conversion after 30 min irradiation.
[b]No reaction.

technique.[41d] The CA-sensitized photoreaction of **77a** in deoxygenated acetone-d_6 exhibits strong nuclear spin polarization effects for **77a** and the CA adducts **84a** and **85a**. The methylene and olefinic protons of **77a** appear as emissions whereas all of the methylene protons of **84a** and **85a** appear as enhanced absorptions. The observed signal direction and polarization-determining parameters suggest that a bisected structure is likely for the trimethylenemethane cation radical **82**$^{•+}$ in which the positive charge is highly localized on the diarylmethylene moiety and the unpaired electron is delocalized over the allyl moiety. A plausible route to form **82**$^{•+}$ is a least motion pathway requiring only rotation of the methylene and not the bulkier diarylmethylene group of **77**$^{•+}$. Similarly, CIDNP monitoring of the photoreaction of **83a** shows a strong emission for the cyclopropane singlet and no polarizations for the aromatic protons. The observed spin polarization effects for **83a** suggest that **83**$^{•+}$ is essentially the 1,1-diarylethylene cation radical in which the unpaired electron and positive charge are located predominantly in the π system. As the *ab initio* calculations of Du and Borden[44] and the MNDO/UHF calculations of C-2–C-3 Kikuchi and co-workers[45] suggest, bond cleavage of the σ-type cation radical **77**$^{•+}$ takes place exothermically while that of the π-type cation radical **83**$^{•+}$ does not.

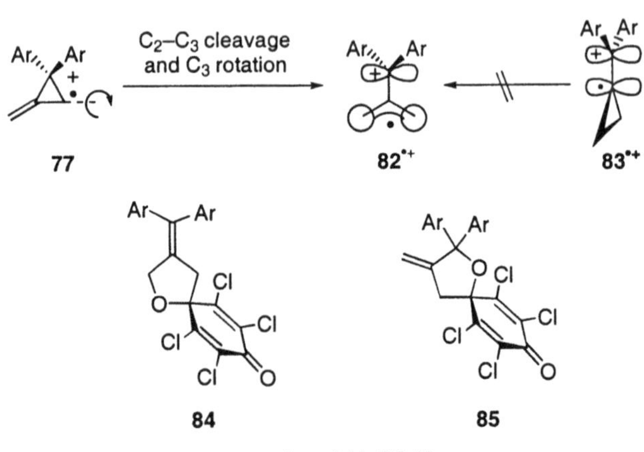

Scheme 21.

Nanosecond laser flash photolyses of **77a** under TCNB- and $NMQ^+BF_4^-$-sensitized conditions provide strong spectroscopic evidence not only for the intermediacy of cation radical **82a**$^{\bullet+}$ in the SET-sensitized photoreactions but also for the trimethylenemethane biradical intermediate **82a**$^{\bullet}$.[41c] Laser excitation (308 nm) of TCNB with **77a** in acetonitrile gives rise to two intense transient absorptions with maxima at 350 and 500 nm whereas in dichloromethane only the shorter wavelength transient absorption is observed. In contrast, under the $NMQ^+BF_4^-$-sensitized conditions in acetonitrile the longer wavelength transient absorption predominates but the shorter wavelength band is extremely weak (Table 5).

The mechanistic connection which exists between the rearrangement, oxygenation, and transient absorption data provides strong evidence for the participation of **82**$^{\bullet+}$ and **82**$^{\bullet}$ in the degenerate rearrangement pathway. Both the degenerate rearrangement and oxygenation reactions take place efficiently under the SET-sensitized conditions in which two transient absorption bands are observed (Table 5, entries 5 and 7). Because degenerate rearrangement of **77a** efficiently occurs under the conditions which give rise to the shorter wavelength absorption only (Table 5, entry 6), the 350-nm transient is assigned to the intermediate in the degenerate rearrangement process. On the other hand, the 500-nm transient is assigned to the precursor of the oxygenation products because oxygenation takes place efficiently under the SET-sensitized conditions in which the longer wavelength transient absorption band predominates (Table 5, entries 4 and 8). By comparing the λ_{max} of cation **86a**$^+$ and radical **86a**$^{\bullet}$, we can assign the 500- and 350-nm transients to **82a**$^{\bullet+}$ and **82a**$^{\bullet}$, respectively. Thus, the maxima at 500 and 350 nm are ascribed to the respective 1,1-dianisylethyl cation and 1,1-dianisylethyl radical moieties of the bisected intermediates **82a**$^{\bullet+}$ and **82a**$^{\bullet}$.

Although EPR signals related to hydrocarbon cations radicals generated by electrochemical oxidation or chemical oxidation can be readily detected, only a few examples[46] have been reported for cation radicals that are produced by irradiation of solutions of electron donors and an acceptor. Because electron spin polarization offers the advantage of detecting transient species via their EPR signal intensities, chemically induced dynamic electron polarization (CIDEP) spectra can give information not only about short-lived radical intermediates

***Table 5*.** Photostationary Ratios (d_2-**77a**:d_2-**77a′**) in the PET Reactions of d_2-**77a**,[a] Yields of **78a** in the PET Photooxygenation of **77a**,[b] and Transient Absorption Maxima (λ_{max}) Observed in Laser Flash Photolysis of **77a**[c] under Various PET Conditions

Entry	*Conditions*	d_2-**77a**:d_2-**77a′** *(Time, h)*	*Yield of* **78a***(%) (time, min)*	λ_{max} *(nm)* **82a**	λ_{max} *(nm)* **82a**$^{\cdot+}$	ΔOD(**82a**$^{\cdot+}$)/ΔOD(**82a**$^{\cdot}$)[d]
1	DCA/CH_3CN	58:42 (4.5)	100 (15)	e	f	
2	DCA/CH_2Cl_2	56:44 (2)	19 (15)	e	f	
3	DCA/C_6H_6	57:43 (3)	3(120)	e	f	
4	DCA–biphenyl/CH_3CN	slow	100 (15)	e	494	
5	TCNB/CH_3CN	54:46 (4.5)	100 (20)	351	500	1.3
6	TCNB/CH_2Cl_2	56:44 (3)	4 (30)	354	f	~0
7	TCNB–biphenyl/CH_2Cl_2	56:44 (2)	96 (15)	354	508	2
8	$NMQ^+BF_4^-$–toluene/CH_3CN	slow	100 (5)	350[g]	498[g]	>10

Notes: [a]Under N_2. [d_2-**77a**] = 100 mM. Deuterated solvent and cosensitizer were used.

[b]Under O_2. [**77a**] = 10 mM.

[c]Under N_2. [**77a**] = 1 mM.

[d]Ratio of ΔOD of 82a$^{\cdot+}$ to that of **82a**$^{\cdot}$ at 200 ns after excitation.

[e]Not observable.

[f]No transient absorption was observed.

[g]Under air.

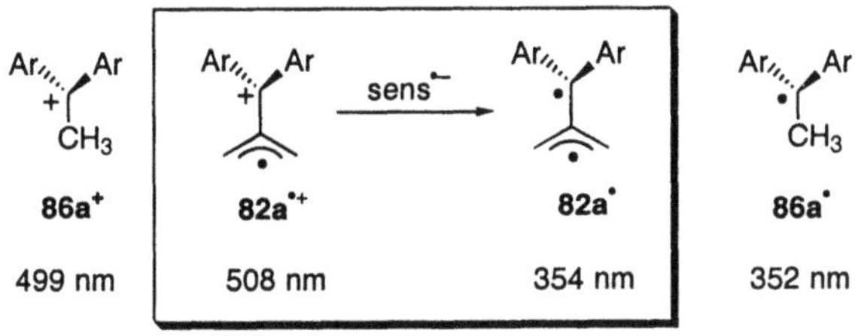

a: Ar = 4-MeOC$_6$H$_4$, λ_{max} in CH$_2$Cl$_2$

Scheme 22.

but also about the nature of the excited states which take part in the reaction. The excited triplet states of quinones gain strong polarization during anisotropic intersystem crossing and transfer this polarization to resulting radical intermediates, if ensuing reactions occur faster than spin lattice relaxation of the excited triplet states. Using this technique, Ishiguro and co-workers[46a] recently succeeded in detecting the quadricyclane cation radical formed in a CA-sensitized electron-transfer photoreaction in polar solvent at low temperature.

We have applied the time-resolved EPR technique to clarify the sequence followed in the SET-photoinduced degenerate methylenecyclopropane rearrangement of d_2-**77a** and have succeeded in confirming the participation of a trimethylenemethane cation radical intermediate. The CIDEP spectrum of **82a**$^{•+}$ was obtained at a delay time of 1 μs after the laser pulse (355 nm) under CA-photosensitized conditions in dimethylsulfoxide at ambient temperature. The E*/E spectrum is easily reproduced by the superposition of the TM (E) and RPM (E/A) signals,[47] where E and A denote emission and enhanced absorption of the microwave radiation, respectively. The CA anion radical appears as an intense signal at 343.135 mT (g = 2.0058).[48] The well-separated five-line pattern is analyzed with two hyperfine splitting (hfs) constants corresponding to the cation radical **82a**$^{•+}$ (g = 2.0031). As the hfs constants and g value shown in Scheme 23 are close to those of the parent allyl radical,[49] the unpaired electron appears to be mainly distributed over the allyl moiety and the positive charge is localized on the dianisylmethyl moiety of **82a**$^{•+}$, thus indicating its bisected structure. The bisected structure of **82a**$^{•+}$ deduced by EPR spectroscopy is consistent with that assigned by use of laser flash photolysis and CIDNP methodologies.

On the other hand, irradiation (355 nm) of AQ and **77a** in a dichloromethane matrix at 22 K gives rise to a characteristic EPR spectrum of randomly oriented triplet species ascribed to **82a**$^{\bullet}$ along with **82a**$^{\bullet+}$. In addition to the $|\Delta Ms| = 1$ transition signals, a weak $|\Delta Ms| = 2$ transition is observed at 167.3 mT. The triplet EPR signal of **82a**$^{\bullet}$ persists at cryogenic temperature and the Curie plot of $|\Delta Ms| = 2$ transition signal intensity gives a straight line between 4.2 and 40 K, indicating that the ground state of **82a**$^{\bullet}$ is triplet, which is normal for trimethylenemethane biradicals.[19,20,30] Analysis of the spectrum gives rise to estimates of the zero-field splitting (zfs) parameters for **82a**$^{\bullet}$ (Table 6) together with those of the parent[50] and diphenyl-substituted cyclic trimethylenemethane biradicals.[51] The zfs parameters of **82a**$^{\bullet}$ are characterized by small $|D/hc|$ and relatively large $|E/hc|$ values as compared with those of other diphenyl-substituted trimethylene-methane derivatives.

Because the delocalization of the unpaired electron may not be changed significantly by the substitution of the methoxy group on the phenyl ring, the reduction of the $|D/hc|$ value of **82a**$^{\bullet}$ reflects the change in the molecular geometry. Changes in the zfs associated with deviations in planar conformations were previously explained by use of simple Hückel MO calculations on a series of biphenyl derivatives,[52] conjugated enones,[53] and conjugated trimethylenemethane[54] biradicals. The results indicate that distortion causes a reduction of the $|D/hc|$ value in all cases. Because **82a**$^{\bullet}$ is generated by the back electron transfer to the bisected cation radical **82a**$^{\bullet+}$, **82a**$^{\bullet}$ must be the bisected

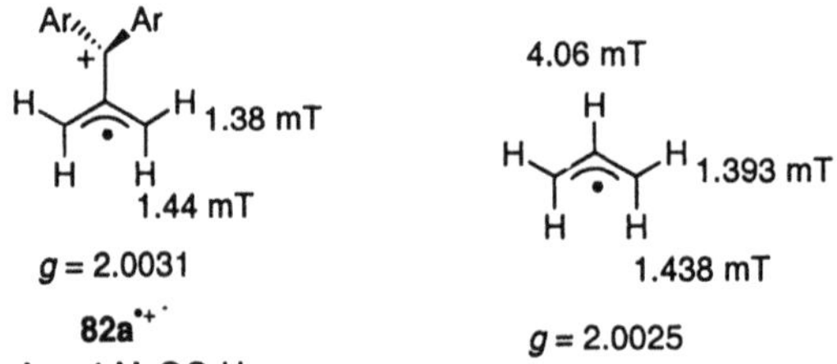

82a$^{\bullet+}$
a: Ar = 4-MeOC$_6$H$_4$

Scheme 23.

species. The small |*D/hc*| value is rationalized by the contribution of the π-orbital rotation around the C=C double bond. This conclusion is also supported by the relatively large |*E/hc*| value. As the |*E/hc*| value relates to the molecular symmetry of the triplet, the value increases with an increase in the twisting angle of the C=C double bond from the molecular plane of the allyl moiety.

Energetics and Mechanisms of Photoinduced Electron Transfer Degenerate Methylenecyclopropane Rearrangements

The accumulated spectroscopic evidence clearly points to the participation of both $\mathbf{82a}^{\bullet+}$ and $\mathbf{82a}^{\bullet}$ in the SET-photoinduced degenerate methylenecyclopropane rearrangement of d_2-**77a**. When combined with the chemical results, it is possible to propose a cation radical cleavage (CRCL)–diradical cyclization (DRCY) mechanism for this process (Scheme 24). Cation radicals d_2-$\mathbf{77a}^{\bullet+}$ and d_2-$\mathbf{77a}'^{\bullet+}$ undergo the C-2–C-3 bond cleavage to form d_2-$\mathbf{82a}^{\bullet+}$ which then reacts exothermically. Because recyclization of d_2-$\mathbf{82a}^{\bullet+}$ to d_2-$\mathbf{77a}^{\bullet+}$ and d_2-$\mathbf{77a}'^{\bullet+}$ is endothermic, competitive back electron transfer from a sensitizer anion radical to d_2-$\mathbf{82a}^{\bullet+}$ operates efficiently to form d_2-$\mathbf{82a}^{\bullet}$, which, in turn, undergoes cyclization to reorganize d_2-**77a** and d_2-**77a′**.

This mechanism has been verified by employment of time-resolved photoacoustic calorimetric experiments.[41c] From measurements with

Table 6. Zero-Field Splitting Values for the Parent and Aryl-Substituted Trimethylenemethane Diradicals

	 [a]	 [b] **82a•**	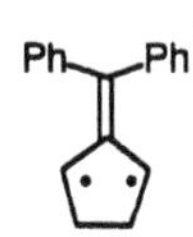 [c]	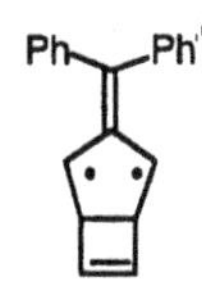 [d]
|*D/hc*| (cm^{-1})	0.024	0.0116	0.0180	0.0174
|*E/hc*| (cm^{-1})	0	0.0038	0.0013	0.0006

Notes: [a]Ref. 50.
[b]Ar = 4-MeOC$_6$H$_4$.
[c]Ref. 51a.
[d]Ref. 51b.

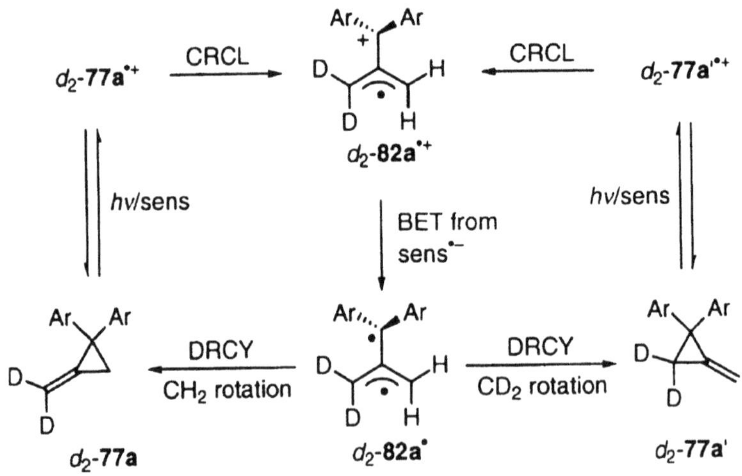

Scheme 24.

the **77a**–DCA–biphenyl system, the energy of the ion radical pair, ΔH_{irp}([**82a**$^{•+}$/DCA$^{•-}$]), is determined to be 37 ± 0.8 kcal/mol. With the energy of ΔH_{irp}([**77a**$^{•+}$/DCA$^{•-}$]) calculated to be 53 kcal/mol, recyclization of **82a**$^{•+}$ to **77a**$^{•+}$ is at least 18 kcal/mol endothermic. MNDO/UHF calculations of Kikuchi and co-workers[45] support the endothermicity of this direct recyclization at the cation radical stage. In contrast, back electron transfer from DCA$^{•-}$ to **82a**$^{•+}$ is estimated to be 20.5 kcal/mol exothermic by using the oxidation potential of the 1,1-dianisylethyl radical (**86a**$^{•}$) (–0.06 V vs SCE in CH_3CN), determined by photomodulation voltammetry, and the reduction potential of DCA (–0.95 V vs SCE in CH_3CN). The highly exothermic back electron transfer, thus, operates to form d_2-**82a**$^{•}$ which is estimated to be about 16.5 kcal/mol higher in energy than d_2-**77a** or d_2-**77a′**.

The most characteristic feature of the photoinduced electron transfer degenerate methylenecyclopropane rearrangement is the intervention of both cation radical and biradical intermediates. In this case, highly endothermic recyclization of d_2-**82a**$^{•+}$ to d_2-**77a**$^{•+}$ and d_2-**77a′**$^{•+}$ allows the operation of the important back electron transfer. Nevertheless, such a back electron transfer process may operate generally if a highly stabilized cation radical intermediate is formed in a highly exothermic process. In fact, the photoinduced electron transfer degenerate Cope rearrangement of 2,5-diaryl-3,3,4,4-tetradeuterio-1,5-hexadiene (d_4-**87**) proceeds in a similar manner.[55] Here, highly exothermic cyclization of d_4-**87**$^{•+}$ forms the highly stabilized cation radical intermediate d_4-**88**$^{•+}$, which is subsequently reduced to d_4-**88**$^{•}$ by a sensitizer anion radical (Scheme 25). The reorganization of d_4-**87** and d_4-87′ occurs through d_4-**88**$^{•}$.

3.2. The Photoinduced Electron Transfer Methylenecyclopropane Rearrangement of Methylenespiropentanes and Cyclopropylidenecyclopropanes

Because both 1-methylenespiropentane (**89**) and 1-cyclopropylidenecyclopropane (**95**) include methylenecyclopropane moiety in their structures, the methylenecyclopropane-type reversible interconversion between **89** and **95** is expected to occur upon pyrolysis, involving a trimethylenemethane biradical intermediate. However, such a rearrangement formally does not take place, though **95** rearranges to **89**.[56] Instead, on pyrolysis at 320 °C, **89** rearranges to dimethylenecyclobutanes (**91** and **93**) through the tetramethylene-ethane biradical **90**$^{•}$ and the vinylic–allylic biradical **92**$^{•}$, respectively.[57] Presumably, biradical **94** formed by C-2–C-5 bond fission

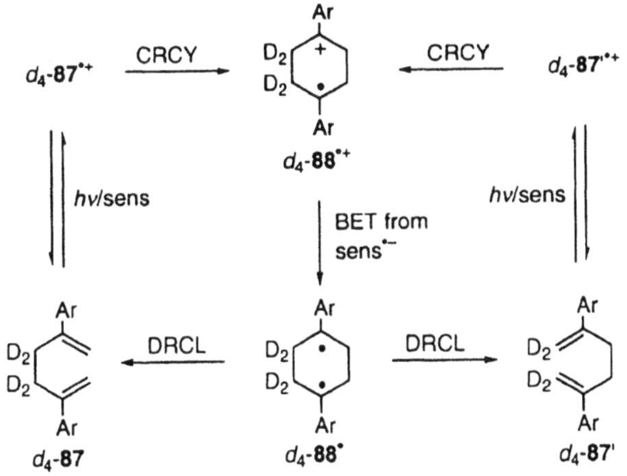

Scheme 25.

undergoes subsequent cyclopropyl–allyl rearrangement to form **90$^{\bullet}$** whereas the vinylic–cyclopropyl biradical formed by C-1–C-5 bond fission rearranges to **92$^{\bullet}$**.

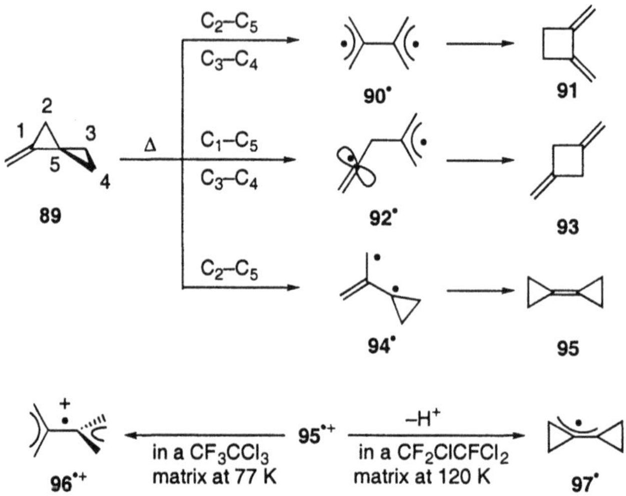

Scheme 26.

Although electron-transfer-promoted reactions of **89** are unknown, cation radical **95$^{\bullet+}$** has been generated by γ-irradiation and its matrix-dependent reactivity was recently reported by Gerson and his co-workers.[58] In a CF_3CCl_3 matrix at 77 K, **95$^{\bullet+}$** rearranges rapidly to the bisected tetramethyleneethane cation radical **96$^{\bullet+}$**. However, deprotonation occurs at 120 K in a $CF_2ClCFCl_2$ matrix to form **97$^{\bullet}$** (Scheme 26). It is reasonable to assume that cation radical **96$^{\bullet+}$** is formed by the cyclopropyl–allyl rearrangement of **94$^{\bullet+}$** itself generated from **95$^{\bullet+}$**.

Unlike the thermal rearrangement of **89** and the cation radical rearrangement of **95$^{\bullet+}$**, the 2,2-diaryl-substituted derivatives of **89** and **95** undergo reversible methylenecyclopropane rearrangements under photoinduced electron transfer conditions.[26] Like 1-methylenecyclopropanes (**77a–c**), 2,2-diaryl-1-methylenespiropentanes (**98a–c**) are good electron donors and their oxidation potentials are low enough to lead to quenching of the excited singlet state of DCA at a near diffusion-controlled rate in acetonitrile. On irradiation of a solution of DCA and **98a** under nitrogen at 10 °C, a photostationary mixture of **98a** and **99a** is formed in the ratio of 1.5:1 and in good yield. The observations that similar electron transfer photoreactions occur to produce **100a** when oxygen-saturated solutions of these substances are irradiated and that **102a** is not formed in the photostationary mixture suggest that the bisected trimethylenemethane cation radical **101a**$^{\cdot+}$ is formed as a common intermediate between **98a**$^{\cdot+}$ and **99a**$^{\cdot+}$.

One intriguing finding in the study of this photoinduced electron transfer process is that the relative ratio of **99** to **98** in the photostationary mixture increases with a decrease in the electron-donating ability of the substrates. Specifically, photostationary ratios of **99**:**98** are 4.5:1 and 6.9:1, respectively, for the tolyl and phenyl derivatives. One possible reason for this phenomenon may be that the π-type nature of cation radical **99**$^{\cdot+}$ increases with a decrease in the electron-donating strength of the aryl group of **99**. A cation radical of this type (like **83**$^{\cdot+}$) would be reluctant to undergo ring cleavage. As expected,

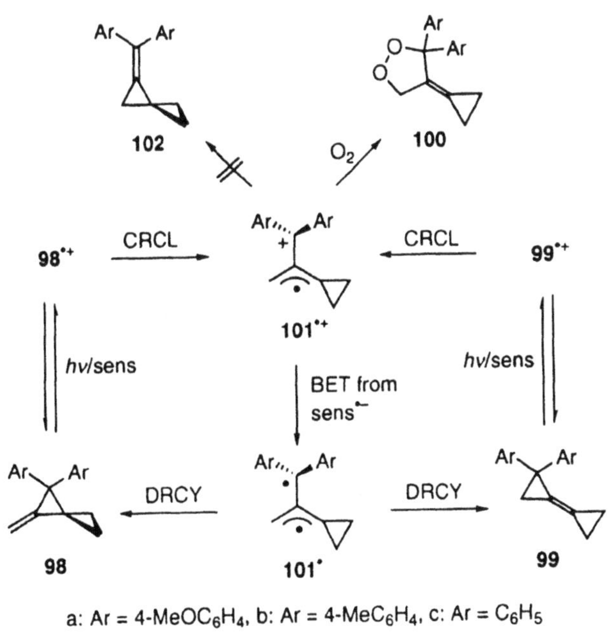

Scheme 27.

99a is thermally labile and rearranges to **98a** on heating at 80 °C. Thus, the photoinduced electron transfer reaction can serve as a useful procedure to prepare thermally labile cyclopropylidenecyclopropane derivatives. It is noteworthy that the cyclopropyl radical function in $\mathbf{101}^{\cdot+}$ does not undergo an exothermic cyclopropyl–allyl rearrangement to form the tetramethyleneethane cation radical under photoinduced electron transfer conditions. These results are essentially the same as those from studies of the electron transfer photoreaction of **77**. The rearrangement sequence presumably is the same as the one proposed for degenerate methylenecyclopropane rearrangement of d_2-**77** as shown in Scheme 27. Preliminary findings in our recent spectroscopic studies support the proposed rearrangement sequence.[59]

ACKNOWLEDGMENTS

We thank many graduate and undergraduate students and Dr. Masaki Kamata (Niigata University) for their collaboration with and contribution to studies from our laboratory which are summarized herein. We acknowledge Professor Joshua L. Goodman (University of Rochester), Dr. Takanori Suzuki (Hokkaido University), and Dr. Danial D. M. Wayner (National Research Council of Canada) for their collaboration in photoacoustic calorimetric, X-ray crystallographic, and photomodulation voltammetric investigations, respectively. Acknowledgment is also made for financial support provided by the Ministry of Education, Science, Sports, and Culture, Japan.

REFERENCES

1. Roth, H. D. In *Topics in Current Chemistry: Photoinduced Electron Transfer IV*; Mattay, J., Ed.; Springer-Verlag: Berlin, 1992; Vol. 163, pp. 131–245.
2. (a) Du, P.; Hrovat, D. A.; Borden, W. T. *J. Am. Chem. Soc.* **1988**, *110*, 3405–3412. (b) Collins, J. R.; Gallup, G. A. *J. Am. Chem. Soc.* **1982**, *104*, 1530–1533. (c) Wayner, D. D. M.; Boyd, R. J.; Arnold, D. R. *Can. J. Chem.* **1985**, *63*, 3283–3289. (d) Haselbach, E. *Chem. Phys. Lett.* **1970**, *7*, 428–430. (e) Rowland, C. G. *Chem. Phys. Lett.* **1971**, *9*, 169–173. (f) Wayner, D. D. M.; Boyd, R. J.; Arnold, D. R. *Can. J. Chem.* **1983**, *61*, 2310–2315.
3. EPR: (a) Shida, T.; Takumura, Y. *Radiat. Phys. Chem.* **1983**, *21*, 157. (b) Qin, X.-Z.; Williams, F. *Chem. Phys. Lett.* **1984**, *112*, 79–83. (c) Qin, X.-Z.; Snow, L. D.; Williams, F. *J. Am. Chem. Soc.* **1984**, *106*, 7640–7641. (d) Qin, X.-Z.; Williams, F. *Tetrahedron* **1986**, *42*, 6301–6314.
4. CIDNP: Roth, H. D.; Schilling, M. L. M.; Schilling, F. C. *J. Am. Chem. Soc.* **1985**, *107*, 4152–4158. See also Ref. 1 and references cited therein.

5. (a) Rao, V. R.; Hixson, S. S. *J. Am. Chem. Soc.* **1979**, *101*, 6458–6459. (b) Dinnocenzo, J. P.; Zuilhof, H.; Lieberman, D. R.; Simpson, T. R.; McKechney, M. W. *J. Am. Chem. Soc.* **1997**, *119*, 994–1004. (c) Dinnocenzo, J. P.; Simpson, T. R.; Zuilhof, H.; Todd, W. P.; Heinrich, T. *J. Am. Chem. Soc.* **1997**, *119*, 987–993.
6. Somich, C.; Mazzocchi, P. H.; Edwards, M.; Morgan, T.; Ammon, H. L. *J. Org. Chem.* **1990**, *55*, 2624–2630.
7. (a) Takahashi, Y.; Nishioka, N.; Endoh, F.; Ikeda, H.; Miyashi, T. *Tetrahedron Lett.* **1996**, *37*, 1841–1844. (b) Takahashi, Y.; Ohaku, H.; Nishioka, N.; Ikeda, H.; Miyashi, T.; Gormin, D. A.; Hilinski, E. F. *J. Chem. Soc. Perkin Trans. 2* **1997**, 303–308. (c) Takahashi, Y.; Endoh, F.; Ohaku, H.; Wakamatsu, K.; Miyashi, T. *J. Chem. Soc. Chem. Commun.* **1994**, 1127–1128. (d) Maruyama, K.; Imahori, H.; Ozawa, Y. *Chem. Lett.* **1989**, 2117–2118.
8. Ichinose, N.; Mizuno, K.; Hiromoto, Z.; Otsuji, Y. *Tetrahedron Lett.* **1986**, *27*, 5619–5620.
9. Mizuno, K.; Yoshioka, K.; Otsuji, Y. *Chem. Lett.* **1983**, 941–944.
10. Dinnocenzo, J. P.; Todd, W. P.; Simpson, T. R.; Gould, I. R. *J. Am. Chem. Soc.* **1990**, *112*, 2462–2464.
11. Mizuno, K.; Ichinose, N.; Otsuji, Y. *Chem. Lett.* **1985**, 455–458.
12. Wong, P. C.; Arnold, D. R. *Tetrahedron Lett.* **1979**, 2101–2104.
13. Karki, S. B.; Dinnocenzo, J. P.; Farid, S.; Goodman, J. L.; Gould, I. R.; Zona, T. A. *J. Am. Chem. Soc.* **1997**, *119*, 431–432.
14. Miyashi, T.; Kamata, M.; Mukai, T. *J. Am. Chem. Soc.* **1987**, *109*, 2780–2788.
15. Dinnocenzo, J. P.; Schmittel, M. *J. Am. Chem. Soc.* **1987**, *109*, 1561–1562.
16. (a) Arnold, D. R.; Humphreys, R. W. R. *J. Am. Chem. Soc.* **1979**, *101*, 2743–2744. (b) Gollnick, K.; Xiao, X.-L.; Paulmann, U. *J. Org. Chem.* **1990**, *55*, 5945–5953. (c) Gollnick, K.; Paulmann, U. *J. Org. Chem.* **1990**, *55*, 5954–5966. (d) Schaap, A. P.; Lopez, L.; Anderson, S. D.; Gagnon, S. D. *Tetrahedron Lett.* **1982**, *23*, 5493–5496.
17. (a) Schaap, A. P.; Siddiqui, S.; Prasad, G.; Palomino, E.; Lopez, L. *J. Photochem.* **1984**, *25*, 167–181. (b) Miyashi, T.; Kamata, M.; Mukai, T. *J. Am. Chem. Soc.* **1987**, *109*, 2780–2788. (c) Kamata, M.; Furukawa, H.; Miyashi, T. *Tetrahedron Lett.* **1990**, *31*, 681–684. (d) Shim, S. C.; Lee, H. J. *J. Photochem. Photobiol. A* **1989**, *46*, 59.
18. (a) Miyashi, T.; Takahashi, Y.; Konno, A.; Mukai, T.; Roth, H. D.; Schilling, M. L.; Abelt, C. J. *J. Org. Chem.* **1989**, *54*, 1445–1447. (b) Roth, H. D.; Schilling, M. L.; Abelt, C. J.; Miyashi, T.; Takahashi, Y.; Konno, A.; Mukai, T. *J. Am. Chem. Soc.* **1988**, *110*, 5130–5136.
19. Berson, J. A. In *Rearrangements in Ground and Excited States*; Mayo, P. de, Ed.; Academic: New York, 1980; Vol. 1, pp. 311–390, and references cited therein.
20. Gajewski, J. J. In *Hydrocarbon Thermal Isomerizations. Organic Chemistry, A Series of Monographs*; Wasserman, H. H., Ed.; Academic: New York, 1981; Vol. 44, pp. 43–70, and references cited therein.

21. (a) Flowers, M. C.; Frey, H. M. *J. Chem. Soc.* **1961**, 5550–5551. (b) Burkhardt, P. J. *Diss. Abstr.* **1962**, *23*, 1524.
22. Doering, W. von E.; Gilbert, J. C. *Tetrahedron Suppl.* **1966**, *7*, 397–414.
23. Shen, K. K.-w.; Bergman, R. G. *J. Am. Chem. Soc.* **1977**, *99*, 1655–1657.
24. Ushida, K.; Shida, T.; Walton, J. C. *J. Am. Chem. Soc.* **1986**, *108*, 2805–2807.
25. (a) Takahashi, Y.; Ohaku, H.; Morishima, S.-i.; Suzuki, T.; Ikeda, H.; Miyashi, T. *J. Chem. Soc. Perkin Trans. 1* **1996**, 319–325. (b) Takahashi, Y.; Ohaku, H.; Morishima, S.-i.; Suzuki, T.; Miyashi, T. *Tetrahedron Lett.* **1995**, *36*, 5207–5210.
26. Miyashi, T.; Takahashi, Y.; Ohaku, H.; Ikeda, H.; Morishima, S.-i. *Pure Appl. Chem.* **1991**, *63*, 223–230.
27. Miyashi, T.; Takahashi, Y.; Ohaku, H.; Yokogawa, K.; Morishima, S.-i.; Mukai, T. *Tetrahedron Lett.* **1990**, *31*, 2411–2414.
28. Miyashi, T.; Takahashi, Y.; Yokogawa, K.; Mukai, T. *J. Chem. Soc. Chem. Commun.* **1987**, 175–177.
29. (a) Ullman, E. F. *J. Am. Chem. Soc.* **1959**, *81*, 5386–5392. (b) Ullman, E. F. *J. Am. Chem. Soc.* **1960**, *82*, 505–506.
30. For reviews see, e.g., (a) Dougherty, D. A. In *Kinetics and Spectroscopy of Carbenes and Biradicals*; Platz, M. S., Ed.; Plenum: New York, 1990; pp. 117–142 and references cited therein. (b) Dowd, P. *Acc. Chem. Res.* **1972**, *5*, 242–248. (c) Dowd, P.; Chow, M. *Tetrahedron* **1982**, *38*, 799–807. See also Refs. 19 and 20.
31. For example: (a) Feller, D.; Tanaka, K.; Davidson, E. R.; Borden, W. T. *J. Am. Chem. Soc.* **1982**, *104*, 967–972. (b) Borden, W. T.; Iwamura, H.; Berson, J. A. *Acc. Chem. Res.* **1994**, *27*, 109–116.
32. See references cited in Ref. 31b.
33. For reviews see, e.g., (a) Little, R. D. *Chem. Rev.* **1996**, *96*, 93–114. (b) Nakamura, E. In *Organic Synthesis in Japan. Past, Present, and Future*; Noyori, R., Ed.; Kagaku Dojin: Tokyo, 1992; pp. 275–282. (c) Nakamura, E. *J. Synth. Org. Chem. Jpn.* **1994**, *52*, 935–945.
34. For reviews see, e.g., (a) Rajca, A. *Chem. Rev.* **1994**, *94*, 871–893. (b) Dougherty, D. A. *Acc. Chem. Res.* **1991**, *24*, 88–94.
35. Kende, A. S.; Goldschmidt, Z.; Smith, R. F. *J. Am. Chem. Soc.* **1970**, *92*, 7606–7607.
36. Baum, T.; Rossi, A.; Srinivasan, R. *J. Am. Chem. Soc.* **1985**, *107*, 4411–4415.
37. Dowd, P.; Chow, M. *J. Am. Chem. Soc.* **1977**, *99*, 6438–6440. See also Refs. 30b and 30c.
38. Yamaguchi, T.; Irie, M.; Yoshida, H. *Chem. Lett.* **1973**, 975–978.
39. Maier, G.; Jürgen, D.; Tross, R.; Reisenauer, H. P.; Hess, B. A., Jr.; Schaad, L. J. *Chem. Phys.* **1994**, *189*, 383–399.
40. Komaguchi, K.; Shiotani, M.; Lund, A. *Chem. Phys. Lett.* **1997**, *265*, 217–223.
41. (a) Miyashi, T.; Kamata, M.; Mukai, T. *J. Am. Chem. Soc.* **1986**, *108*, 2755–2757 and **1987**, *109*, 2780–2788. (b) Takahashi, Y.; Miyashi, T.; Mukai, T. *J. Am. Chem. Soc.* **1983**, *105*, 6511–6513. (c) Ikeda, H.; Nakamura, T.; Miyashi,

T.; Goodman, J. L.; Akiyama, K.; Tero-Kubota, S.; Houmam, A.; Wayner, D. D. M. *J. Am. Chem. Soc.* **1998**, *120*, 5832–5833. (d) Miyashi, T.; Takahashi, Y.; Mukai, T.; Roth, H. D.; Schilling, M. L. M. *J. Am. Chem. Soc.* **1985**, *107*, 1079–1080.

42. Noyori, R.; Hayashi, N.; Katô, M. *J. Am. Chem. Soc.* **1971**, *93*, 4948–4950.
43. Gilbert, J. C.; Butler, J. R. *J. Am. Chem. Soc.* **1970**, *92*, 2168–2169.
44. Du, P.; Borden, W. T. *J. Am. Chem. Soc.* **1987**, *109*, 5330–5336.
45. Takahashi, O.; Morihashi, K.; Kikuchi, O. *Tetrahedron Lett.* **1990**, *31*, 5175–5178.
46. (a) Ishiguro, K.; Khudyakov, I. V.; McGarry, P. F.; Turro, N. J.; Roth, H. D. *J. Am. Chem. Soc.* **1994**, *116*, 6933–6934. (b) Mattay, J.; Gersdorf, J.; Buchkremer, K. *Chem. Ber.* **1987**, *120*, 307–318. (c) Batchelor, S. N.; Heikkilä, H.; Kay, C. W. M.; McLauchlan, K. A.; Shkrob, I. A. *Chem. Phys.* **1992**, *162*, 29–45.
47. McLauchlan, K. A. In *Modern Pulsed and Continuous-Wave Electron Spin Resonance*; Keva, L.; Bowman, M. K., Eds.; Wiley: New York, 1990; pp. 285–363.
48. Segal, B. G.; Kaplan, M.; Fraenkel, G. K. *J. Chem. Phys.* **1965**, *43*, 4191–4200.
49. (a) Fessenden, R. W.; Schuler, R. H. *J. Chem. Phys.* **1963**, *39*, 2147–2195. (b) Krusic, P. J.; Meakin, P.; Smart, B. E. *J. Am. Chem. Soc.* **1974**, *96*, 6211–6213.
50. Dowd, P. *J. Am. Chem. Soc.* **1966**, *88*, 2587–2589.
51. (a) Platz, M. S.; McBride, J. M.; Little, R. D.; Harrison, J. J.; Shaw, A.; Potter, S. E.; Berson, J. A. *J. Am. Chem. Soc.* **1976**, *98*, 5725–5726. (b) Hirano, T.; Kumagai, T.; Miyashi, T.; Akiyama, K.; Ikegami, Y. *J. Org. Chem.* **1991**, *56*, 1907–1914.
52. Tanigaki, K.; Taguchi, N.; Yagi, M.; Higuchi, J. *Bull. Chem. Soc. Jpn.* **1989**, *62*, 668–673.
53. Yamauchi, S.; Hirota, N.; Higuchi, J. *J. Phys. Chem.* **1988**, *92*, 2129–2133.
54. Bushby, R. J.; Jarecki, C. *Tetrahedron Lett.* **1986**, *27*, 2053–2056.
55. (a) Ikeda, H.; Minegishi, T.; Abe, H.; Konno, A.; Goodman, J. L.; Miyashi, T. *J. Am. Chem. Soc.* **1998**, *120*, 87–95. (b) Miyashi, T.; Konno, A.; Takahashi, Y. *J. Am. Chem. Soc.* **1988**, *110*, 3676–3677.
56. (a) Perchec, P. Le; Conia, J. M. *Tetrahedron Lett.* **1970**, 1587–1588. (b) Dolbier, W. R., Jr.; Alonso, J. H. *J. Am. Chem. Soc.* **1973**, *95*, 4421–4423. (c) Dolbier, W. R., Jr.; Akiba, K.; Riemann, J. M.; Harmon, C. A.; Bertrand, M.; Bezaguet, A.; Santelli, M. *J. Am. Chem. Soc.* **1971**, *93*, 3933–3940.
57. Dolbier, W. R., Jr. *Tetrahedron Lett.* **1968**, 393–396.
58. (a) Gerson, F.; Meijere, A. de; Qin, X.-Z. *J. Am. Chem. Soc.* **1989**, *111*, 1135–1136. (b) Gerson, F.; Schmidlin, R.; Meijere, A. de; Späth, T. *J. Am. Chem. Soc.* **1995**, *117*, 8431–8434.
59. Ikeda, H.; Shiratori, Y.; Miyashi, T. Unpublished results.

ELECTROCHEMISTRY APPLIED TO THE SYNTHESIS OF FLUORINATED ORGANIC SUBSTANCES

Toshio Fuchigami

Advances in Electron Transfer Chemistry
Volume 6, pages 41–130.

ISBN: 0-7623-0213-5

1. INTRODUCTION

Elemental fluorine was first prepared by Mossian more than a century ago. Among halogens, fluorine is quite characteristic and specific since it has the largest electronegativity (4.0 vs. 3.5 for oxygen) and the sterically second smallest van der Waals radius (1.35 vs. 1.20 Å for hydrogen). A carbon–fluorine bond is also stronger than a carbon–hydrogen bond (485 vs. 414 kJ/mol). In addition, fluorine can participate in hydrogen bonding interactions as an electron pair donor. Therefore, fluoro-organic compounds have unique chemical, physical, and biological properties. Recently, there has been increasing interest in the chemistry of fluoro-organic compounds, which has wide application in various fields such as material science, medicinal chemistry, and theoretical chemistry. However, fluorinated organic compounds are generally not naturally occurring. Therefore, the fluorine atom(s) must be introduced by synthesis into basic organic starting materials which can then be converted to the desired fluorinated products. However, methods for selective fluorination of organic compounds and selective conversion of fluorinated organic compounds are not straightforward because of the reactivity of molecular fluorine and difficulty in applying ordinary synthetic organic methods.

As a consequence of these difficulties, a number of new methods and techniques have been developed for these purposes.[1–10] In recent years, electrochemical electron transfer reactions have been shown to be highly efficient and, consequently, they serve as new tools in fluoro-organic synthesis. However, only a limited number of examples of electrosyntheses of fluoro-organic compounds, except for the well-established anodic perfluorination and anodic trifluoromethylation processes, were reported prior to the 1980s.

This chapter deals with recent advances in the application of electrochemical electron transfer reactions to the synthesis of fluorinated organic substances. The effect of fluorine atoms on the reduction and oxidation potentials of organic compounds is discussed first. Subsequently, recent applications of the electrochemistry to the conversion and functionalization of fluoro-organics (building-block approach) are described. Finally, methods for selective electrochemical fluorination of organic molecules (direct fluorination approach) are briefly considered.

2. CATHODIC REDUCTION OF C–F BONDS

The ease of reduction of a carbon–halogen bond decreases in the order I > Br > Cl > F. The carbon-fluorine bond is the most difficult to reduce because of its large electronegativity. In general, simple alkyl fluorides are not electrochemically reducible. However, some uniquely substituted fluorides can be reduced, as is seen with organofluorides bearing electron-withdrawing groups like $PhCOCH_2F$, CF_3COOEt ($E_{p1/2}$ = −2.36 V vs. SCE in DMF, Hg cathode),[11] and $CF_3CONHPh$ (E_p = −2.32 V vs. −1.90 V for CCl_3CN in MeCN, Hg cathode).[12] Trifluoromethylbenzenes and in particular those bearing electron-withdrawing ester or cyano groups at the *para* position are readily reduced (Eq. 1).[13]

$$p\text{-}XC_6H_4CF_3 \xrightarrow[\text{Pb cathode}]{6e,\ 3H^+ \quad -3F^-} p\text{-}XC_6H_4CH_3 \qquad (1)$$

X = CN (-1.8 V): 58%
= COOMe (-2.0 V): 60%

It has been reported recently that cathodic reduction of trifluoromethyl ketones **1** and imines **3** in the presence of trimethylsilyl chloride (TMSCl) gives β,β-difluoroenolsilyl ethers **2** and enamines **4**, respectively, in good yields (Eqs. 2 and 3).[14] In these reactions, eliminated fluoride ions react with both TMSCl and the products. Therefore, a 3-equivalent excess of TMSCl is necessary for trapping of fluoride ions which are cathodically generated.

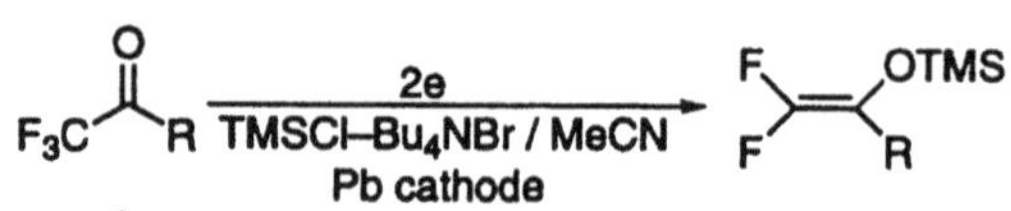

1 **2** (2)

R = Ph 80%
2-Furyl 55%
CH_2CO_2Me 50%

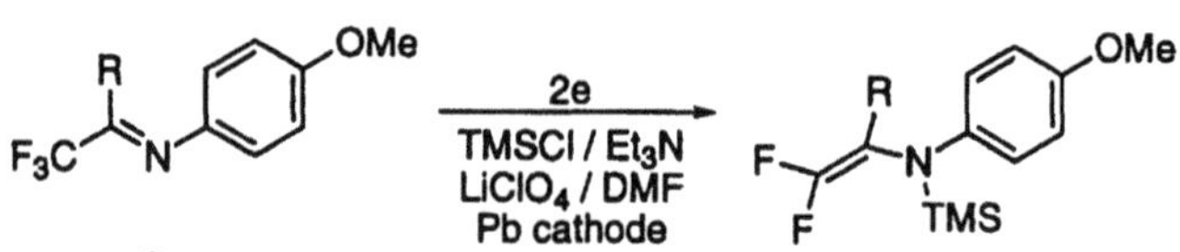

3 **4** (3)

R = CO_2Et 78%
Ph 75%
2-Furyl 57%

On the other hand, poly(tetrafluoroethylene) (PTFE) can be electrochemically reduced.[15] Recently, electrochemical reduction of saturated perfluoroalkanes has been observed on an analytical scale and it was found that the reduction potentials of perfluorocycloalkanes are only slightly more negative than those of the corresponding perfluoroaromatics: perfluorodecaline (E_p = −2.60 V vs Ag/0.01 M $AgClO_4$) versus octafluoronaphthalene (−2.58 V); perfluoromethylcyclohexane (−2.9 V) versus perfluorotoluene (−2.75 V).[16]

Selective defluorination of 1,3-difluorobenzene to fluorobenzene has been successfully carried out by use of cathodic reduction at mercury in diglyme containing Bu_4NBF_4 and a catalytic amount of a dimethylpyrrolidinium (DMP^+) salt.[17] In this reaction, DMP^+ is first reduced to form an amalgam, which reduces difluorobenzene catalytically as shown in Scheme 1. Also, cathodic reduction of perfluorobenzene at an aluminum cathode in aqueous DMF provides benzene in moderate yield (Eq. 4).

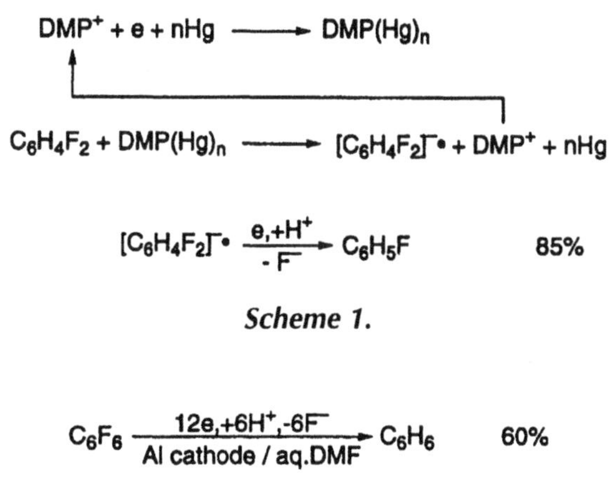

Scheme 1.

$$C_6F_6 \xrightarrow[\text{Al cathode / aq.DMF}]{12e,+6H^+,-6F^-} C_6H_6 \qquad 60\% \qquad (4)$$

3. CATHODIC REDUCTION OF PERFLUORINATED AND POLYFLUORINATED ORGANIC HALIDES

3.1. Direct Reduction

Historically, cathodic reduction was one of the first subjects explored in the area of organofluorine electrochemistry. Perfluoroalkyl halides are reduced cathodically much easier than the corresponding nonfluorinated halides. Rozhkov and his co-workers have investigated the polarographic reduction of perfluoroalkyl halides at a platinum cathode in MeCN.[18] The reduction potentials of these substances, summarized in Table 1, are greatly affected by molecular structure. Accordingly, cyclic halides are much more easily reduced than their acyclic analogues and their reduction potential decreases with increasing length of the perfluoroalkyl (R) chain. The ease of reduction varies in the following order: $(Rf)_3CI$ > $(Rf)_3CBr$ > $(Rf)_2CFI$ > $RfCF_2I$ ~ $(CF_3)_2CFBr$ ~ $(Rf)_3CCl$.

Saveant and his co-workers[19] also measured the reduction potentials of perfluoroalkyl halides by use of cyclic voltammetry. They found that the reduction potentials depend greatly on the nature of the cathode material as shown in Table 2. For example, primary perfluoroalkyl iodides are ca. 0.3 V easier to reduce at mercury than at platinum as a result of the strong interaction of the substrate with mercury. In fact, cathodic reduction of RfI at a mercury cathode provides RfHgI.[20] In contrast to these observations, Ignat'ev et al. have pointed out that a glassy carbon cathode leads to less negative reduction potentials as compared to those measured using a platinum cathode.[21]

Electrochemical reduction of primary perfluoroalkyl halides in the absence of other reagents leads to the formation of mainly monohy-

Table 1. Reduction Potentials of Perfluoroalkyl Halides

RfX	$E_{1/2}$ *V vs SCE*
C_3F_7I	−1.00
$(CF_3)_2CFI$	−0.66
$(CF_3)_3CI$	+0.14
$CF_3CF_2CF_2C(CF_3)_2I$	+0.32
C_4F_9I	−2.30
$(CF_3)_2CFBr$	−1.10
$(CF_3)_3CBr$	−0.14
$(CH_3)_3CBr$	−2.51
$CF_3CF_2CF_2C(CF_3)_2Cl$	−0.97

Note: In 0.1 M Et_4NBF_4/MeCN; Pt cathode.

Table 2. Reduction Potentials (Peak Potentials, E_p vs SCE) of CF_3Br and CF_3I at Various Cathodes[a]

Compound	*Glassy Carbon*	*Pt*	*Au*	*Hg*	*Ni*	*Cu*
CF_3Br[b]	−2.07	−1.55	−1.23	−1.25	−1.33	−1.18
CF_3I[c]	−1.52	−0.95	−0.70	−0.65	—	—

Notes: [a]In 0.1 M BU_4NBF_4/DMF; 0.2 V/s.
[b]At 25 °C.
[c]At 5 °C.

dropolyfluoroalkanes via a pathway involving hydrogen atom or proton abstraction from the solvent. On the other hand, electrochemical reduction of perfluoroalkyl halides in the presence of nucleophiles such as thiolate anions leads to the production of the corresponding perfluoroalkyl sulfides in high yields and with > 100% current efficiency (Eq. 5).[22] The reaction proceeds via SRN_1 mechanism as shown in Scheme 2.

$$\mathrm{RfX + RS^-} \xrightarrow[\text{0.2–0.5 F / mol}]{e} \mathrm{RfSR} \qquad (5)$$

(X = I, Br)

60~85%
Rf = C_nF_{2n+1} (n=1,3,4,6), C_4F_8H
R = C_8H_{17}, Bz, p-ClC_6H_4, p-$CH_3C_6H_4$
63~72%
Rf = $H(CF_2CF_2)_nCH_2$ (n = 1-3)
R = p-ClC_6H_4

Electrocatalytic additions of perfluoroalkyl iodides to 3-hydroxyalkynes have been performed in aqueous KCl emulsions using a carbon fiber cathode. Radical chain mechanisms are responsible for

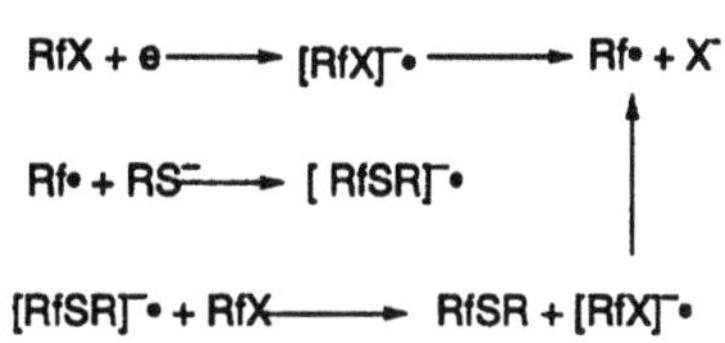

Scheme 2.

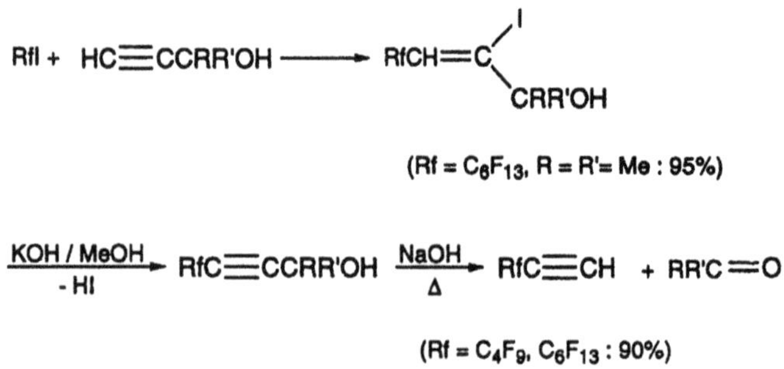

Scheme 3.

these processes. Importantly, the electrolytic products can be readily converted into perfluoroalkyl acetylenes (Scheme 3).[23,24]

Recently, Shono, Kise, and their co-workers have successfully carried out moderately efficient electroreductive coupling reactions of halofluoro compounds with aldehydes in the presence of TMSCl as shown in Eqs. 6–8.[25] Coupling products of type **5** are also cathodically converted to various fluoro compounds as shown in Eq. 9.[25]

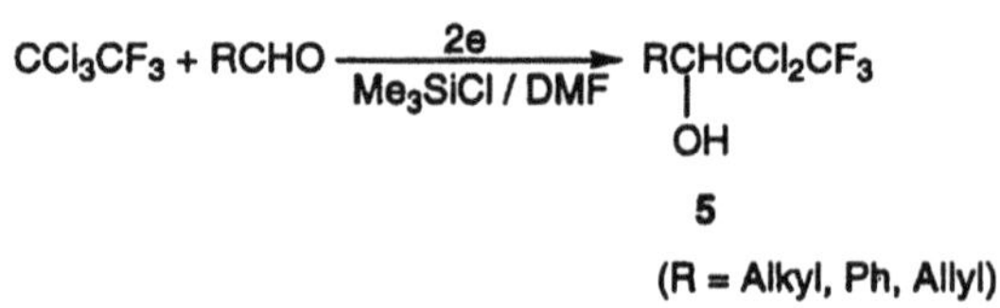

(6)

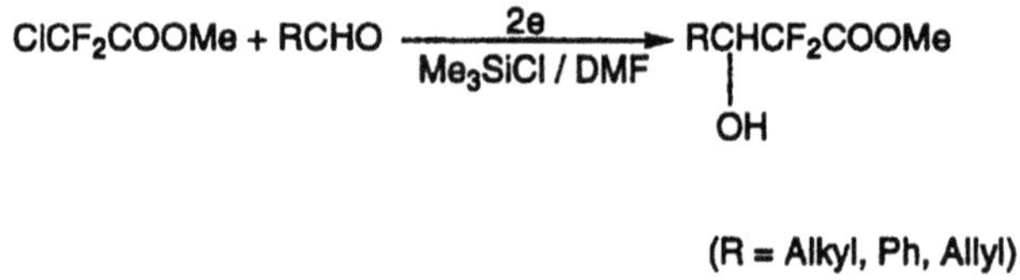

(7)

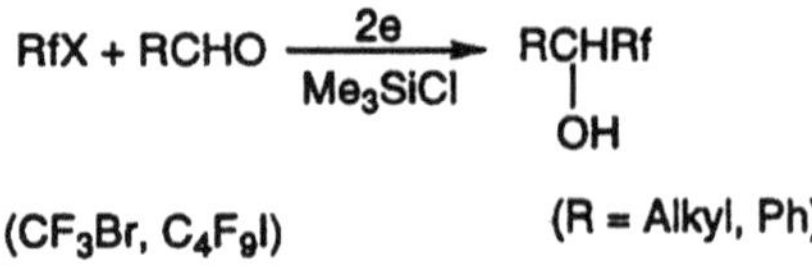

(8)

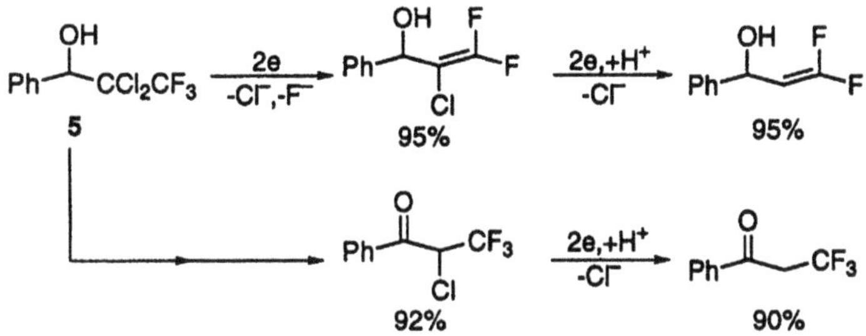
(9)

In contrast, cathodic reduction of dibromodifluoromethane generates difluorocarbene, which is successfully trapped with olefins to yield *gem*-difluorocyclopropane derivatives (Eq. 10).[26]

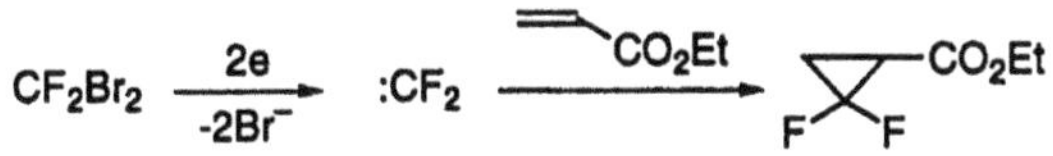
(10)

Perfluoroacyl halides are reduced more easily than the corresponding perfluoroalkyl halides as shown by the data in Table 3.[27] Electrochemical reduction of perfluoroacyl halides results in the formation of dimerization products although the yields of these processes are low (Eq. 11).[27]

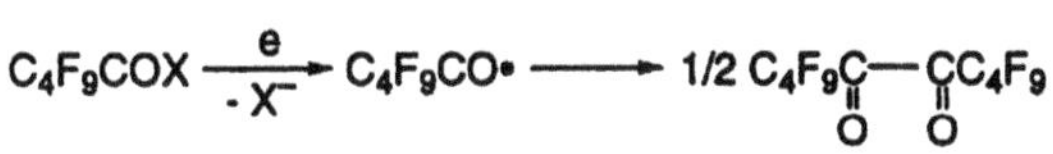
(11)

Table 3. Reduction Potentials of Perfluoroacyl Halides[a] RfCOX

Rf	*X*	*$E_{1/2}$ V vs SCE*
C_4F_9	I	–0.99
C_4F_9	Br	–1.22
C_6F_{13}	Br	–1.13
C_6F_{13}	Cl	–1.28
C_8F_{17}	F	–1.49

Note: [a]In 0.1 M $Bu_4N{\cdot}BF_4/CH_2Cl_2$–MeCN (2:1); glassy carbon cathode.

3.2. Indirect Reduction

In contrast to the direct reduction reactions described above, indirect electrochemical reductions of perfluoroalkyl halides serve as versatile and novel methods for selectively generating perfluoroalkyl radicals. Saveant and his co-workers have offered many interesting examples of reactions of this type. Using terephthalonitrile as a mediator, indirect reduction of CF_3Br in the presence of styrene leads to formation of the dimer of the radical addition product obtained by attack of $CF_3^{\bullet}$ on styrene. On the other hand, when this process is run in the presence of butyl vinyl ether rather than styrene, the mediator reacts with the radical adduct obtained by the attack of $CF_3^{\bullet}$ on the vinyl ether (Scheme 4).[28]

In addition, nitrobenzene redox catalyzed electrolysis of $C_6F_{13}I$ in benzonitrile provides 4-perfluorohexylbenzonitrile as the main product (Scheme 5).[19]

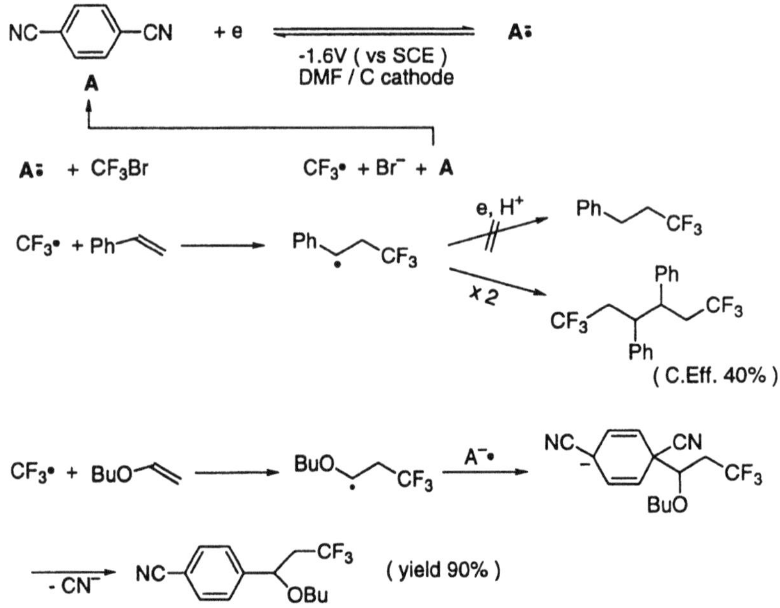

Scheme 4.

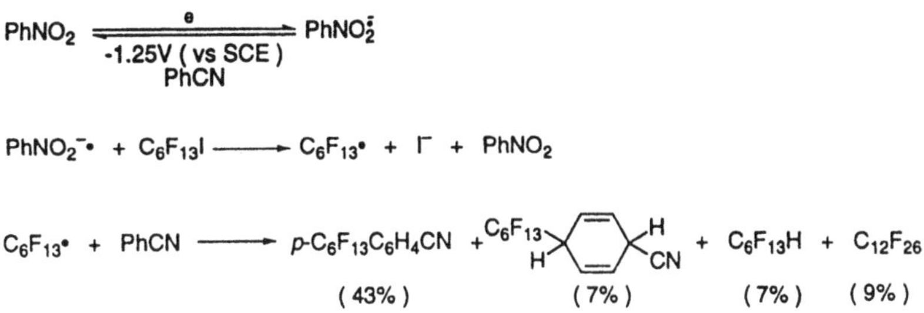

Scheme 5.

Although the reduction of CF_3Br by cathodically generated aromatic anion radicals gives rise to purely catalytic current, cathodically generated $SO_2^{-\bullet}$ does not lead to generation of catalytic currents on reaction with CF_3Br but rather produces trifluoromethyl sulfinate according to an overall two electron per molecule stoichiometry.[29] In the latter case, $SO_2^{-\bullet}$ abstracts a bromine atom from CF_3Br to give the CF_3 radical which further reacts with $SO_2^{-\bullet}$ to give $CF_3SO_2^{-\bullet}$ (Eq. 12).

$$\begin{array}{l} SO_2 \longrightarrow SO_2^{-\bullet} \xrightarrow{CF_3Br} CF_3^{\bullet} + BrSO_2^{-} \\ CF_3^{\bullet} + SO_2^{-\bullet} \longrightarrow CF_3SO_2^{-} \quad \text{(C. Eff. 60\%)} \end{array} \tag{12}$$

Furthermore, Medebielle and Saveant have uncovered elegant examples of electrochemically induced nucleophilic substitution reactions of perfluoroalkyl halides. The reaction mechanism is a slightly modified version of the classical SRN_1 pathway in which initiation occurs by dissociative electron transfer, and the route does not involve the intermediacy of the anion radical of the substrate as shown in Scheme 6.[30,31]

$$\begin{array}{l} M_{ox} + e \rightleftharpoons M_{red} \\ RfX + M_{red} \longrightarrow Rf^{\bullet} + X^{-} + M_{ox} \\ Rf^{\bullet} + Nu^{-} \longrightarrow RfNu^{\bar{\bullet}} \\ RfNu^{\bar{\bullet}} + M_{ox} \longrightarrow RfNu + M_{red} \quad \text{or} \quad RfNu^{\bar{\bullet}} + RfX \longrightarrow RfNu + Rf^{\bullet} + X^{-} \\ \hline RfX + Nu^{-} \longrightarrow RfNu + X^{-} \end{array}$$

Scheme 6.

In this manner, perfluoroalkylated imidazoles are obtained in excellent to good yields (Eq. 13).[30,31] Similarly, indirect cathodic reduction of fluoroalkyl iodide in the presence of various heterocyclic anions such as those of uracil, adenine, and xanthine gives the corresponding fluoroalkylated heterocycles in good yields (Eqs. 14 and 15).[32] A hindered phenolate anion also provides a fluoroalkylated dimeric product in high yield under these reaction conditions (Eq. 16).[31] Furthermore, Medebielle extended this mediator system to chlorodifluoroacetylenic compounds as shown in Eq. 17.[33] This variant seems to be useful for the preparation of biologically interesting cyclic compounds having *gem*-difluoromethylene units.

(13)

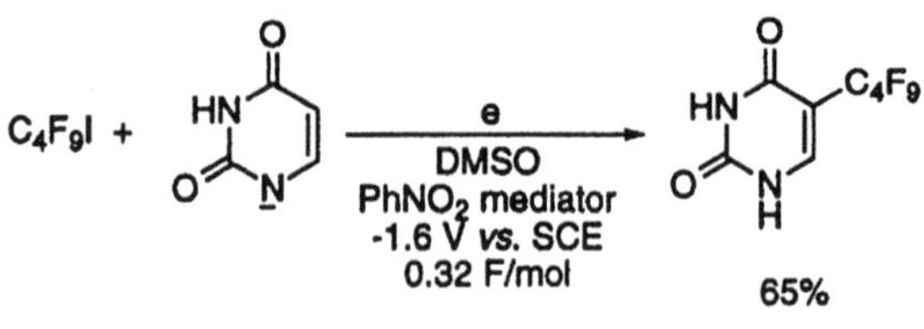

(14)

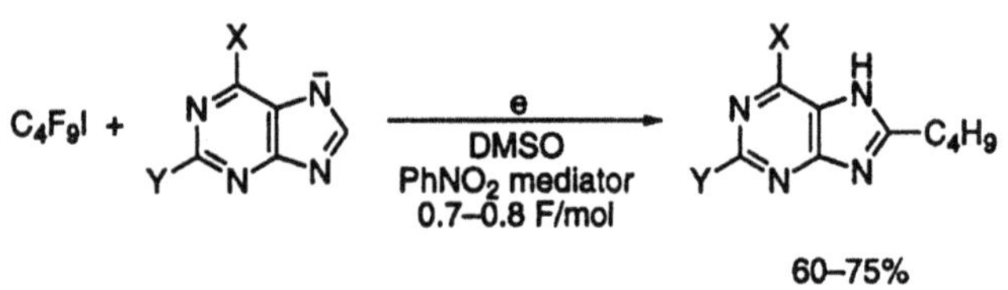

(15)

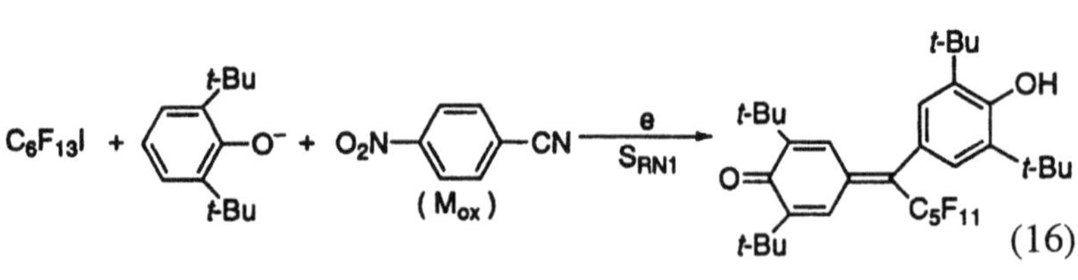

(16)

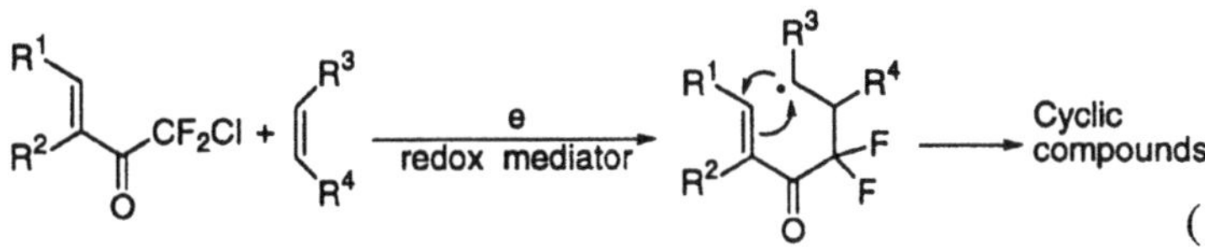

(17)

Ignat'ev and his co-workers found that oxygen can act as mediator for the reduction of perfluoroalkyl halides (Scheme 7).[34] They also showed that the radical anion of oxygen plays the role of a reagent as well as a mediator in this process.[35] Thus, electrochemical reduction of perfluoroalkyl halides in the presence of excess oxygen provides the corresponding perfluorocarboxylic acids in high yields (Eq. 18).

$$O_2 + e \rightleftharpoons O_2^{-\bullet}$$

$$RfX + O_2^{-\bullet} \longrightarrow [RfX]^{-\bullet} + O_2$$

$$[RfX]^{-\bullet} \longrightarrow Rf^{\bullet} + X^-$$

Scheme 7.

$$RfCF_2X + O_2 \xrightarrow{e} RfCOO^- + 2F^- + X^- \quad (18)$$

$Rf = C_2F_5$; $X = I$ 76%
$Rf = C_5F_{11}$; $X = Br$ 67%

Phenylselenate anions generated electroreductively from diphenyldiselenide also can reduce bromodifluoroacetate to form selectively the corresponding radical, which undergoes efficient addition reactions as shown in Eq. 19.[36] α,α-Difluoro-γ-lactams were also prepared using a similar mediator system (Eq. 20).[37]

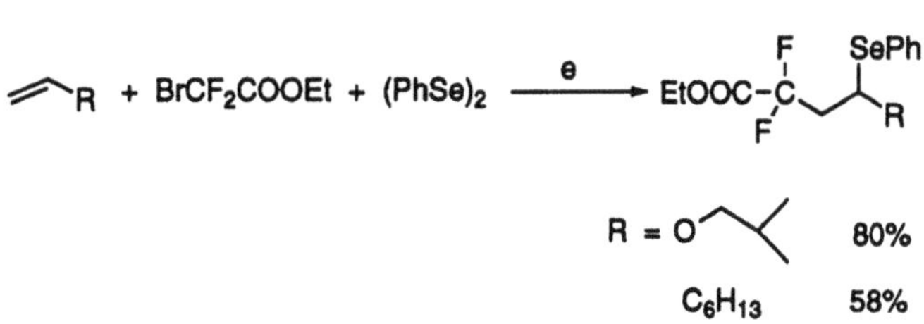

(19)

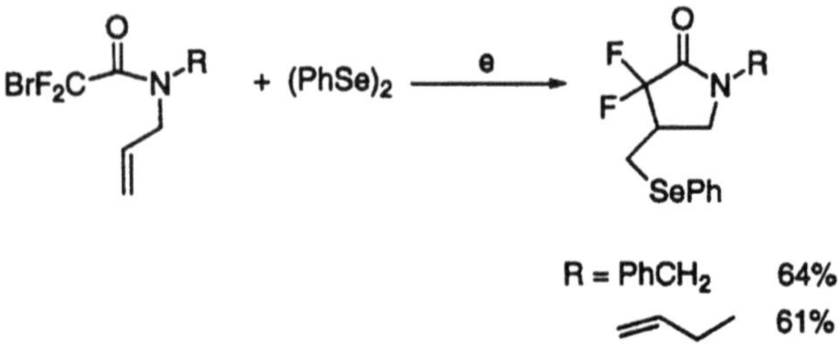

(20)

3.3. Utilization of Sacrificial Anodes

Recently, electrochemical functionalization of organic halides by use of sacrificial anodes has been developed in a remarkable way.[38] The general mechanism for this process is schematically shown for the generation of a divalent cation in Scheme 8. The reaction appears to be similar to that seen in organometallic synthesis, where the reduction is performed by the metal instead of electricity. However, these reactions have been shown to be essentially different from the corresponding organometallic reactions. The electrochemical method has several valuable advantages. For example, since the anode reaction is controlled, an undivided cell can be used. In addition, the reaction occurs in one step and the conditions are quite simple. Sibille et al. have found that the sacrificial zinc anode is highly effective for use in the trifluoromethylation of aldehydes where trifluoromethylated alcohols are generated in almost quantitative yields (Eq. 21).[39] The reaction proceeds via the reduction of Zn(II) salts, followed by a chemical reaction between the reduced metal, CF_3Br, and the aldehyde.

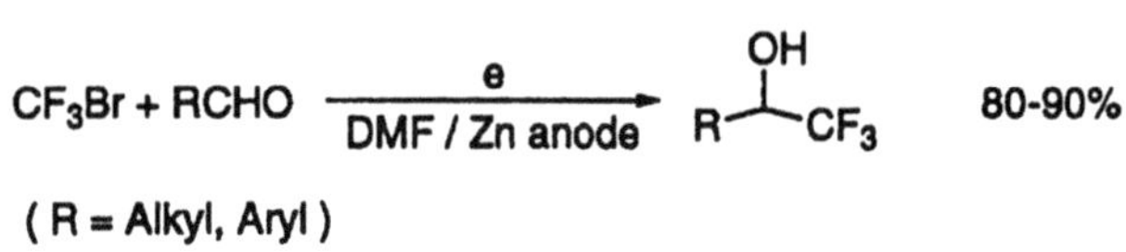

(21)

At the anode : $M \longrightarrow M^{2+} + 2e$

At the cathode : $RX + 2e \longrightarrow R^- + X^-$

In solution : $R^- + E^+ \longrightarrow RE$

Overall reaction : $RX + M + E^+ \xrightarrow{\text{electricity}} RE + MX^+$

Scheme 8.

When ketones are used in this process, the main products are unreactive organozinc species CF_3ZnBr and CF_3ZnCF_3. However, the use of DMF/TMEDA (7:3) as a solvent suppresses the formation of the organozinc species and promotes the carbonyl addition route providing tertiary alcohols in moderate yields (Eq. 22).[40]

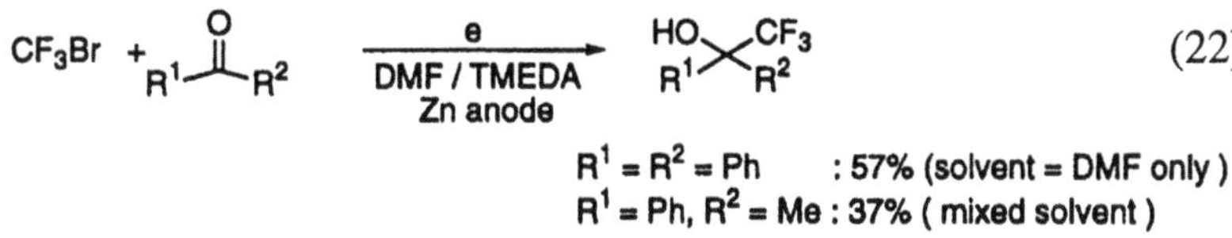

(22)

Formylation can be achieved also when DMF is used as the electrophilic acceptor. Thus, the cathodic reduction of CF_3Br in DMF using an aluminum anode provides trifluoroacetaldehyde in good yield (Eq. 23).[41]

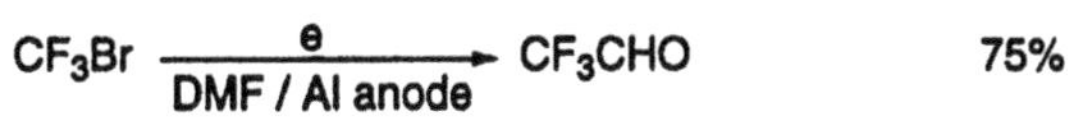

(23)

Even trifluoromethylbenzene can be used as a starting material, which gives various *gem*-difluoro compounds in one step (Eq. 24).[42] It should be noted that the defluorinative-silylation steps can be strictly governed simply by controlling the charge passed and the current density (Eq. 25).[43]

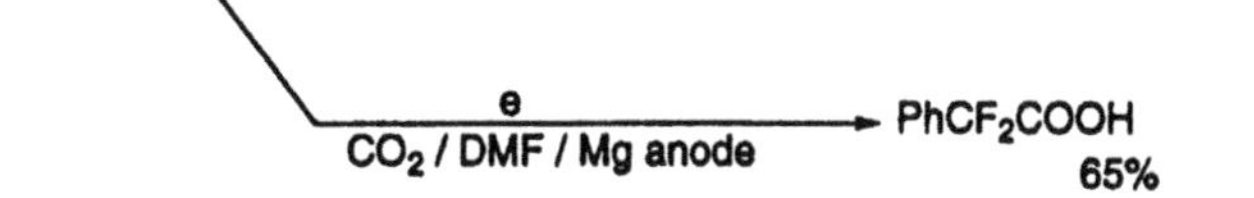

(24)

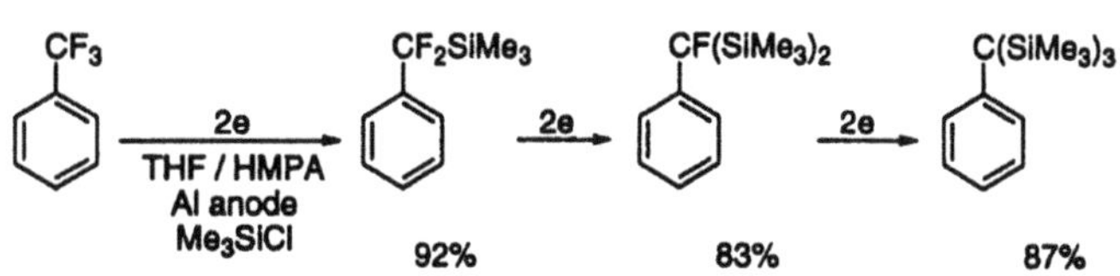

(25)

Trifluoromethyltrimethylsilane is a highly useful trifluoromethylating reagent. Efficient electrochemical trimethyl silylation of bromotrifluoromethane has been developed by Prakash et al. (Eq. 26).[44] A one-pot electrosynthesis of trifluoromethanesulfinic acid has also been achieved at a sacrificial anode with good current efficiency (Eq. 27).[45] Nedelec et al. have achieved the electrochemical cross-coupling of CF_3CCl_3 with $PhCH_2Br$ by using a sacrificial aluminum anode (Eq. 28).[46]

$$CF_3Br + Me_3SiCl \xrightarrow[\text{Al anode}]{2e} CF_3SiMe_3 \qquad (26)$$

$$CF_3Br + SO_2 \xrightarrow[\text{DMF / Zn or Mg anode}]{e} CF_3SO_2^- \quad 60\text{-}70\% \qquad (27)$$

$$CF_3CCl_3 + PhCH_2Br \xrightarrow[\text{THF, Al anode}]{e} CF_3CCl_2CH_2Ph \quad 60\% \qquad (28)$$

The cross-coupling of CF_3Br with aromatic and heteroaromatic halides has also been achieved using a sacrificial copper anode (Eq. 29).[47]

$$CF_3Br + ArX \xrightarrow[\text{DMF / Ph}_3\text{P}]{e} ArCF_3 \qquad (29)$$

ArX = 4-Iodoanisole : 90%
4-Iodonaphthalene : 98%
3-Bromopyridine : 98%
3-Bromoquinoline : 98%
2-Bromothiophene : 88%

Furthermore, 2,2-difluoro-3-hydroxyesters are readily obtained from $ClCF_2COOMe$ and carbonyl compounds by electrolysis in a one-compartment cell using a sacrificial zinc anode and a nickel complex as catalyst.[48] The catalytic cycle for this reaction is shown in Scheme 9 with nickel zinc exchange being a key step. In this process, the CH_2Cl_2/DMF solvent (9:1) system leads to suppression of undesired Claisen condensation and an increase in the yield of 2,2-difluoro-3-hydroxyester formation. It is notable that high yields are obtained even with ketones and enolizable aldehydes, which are not good participants in the Reformatsky reaction alternative for producing these substances.

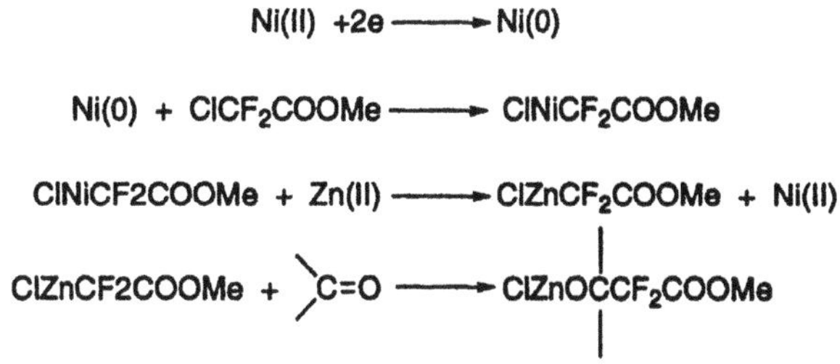

Scheme 9.

In addition, homocoupling of *o*-trifluoromethyl chlorobenzene[49] and *p*-fluorobromobenzene[50] leading to the respective formation of *o,o′*-bis(trifluoromethyl)biphenyl and *p,p′*-difluorobiphenyl[51,52] has been performed using sacrificial anodes.

Quite recently, sulfur-containing heterocycles bearing trifluoromethyl groups have been prepared from 1,1-dibromo-3,3,3-trifluoropropenes by using a sacrificial sulfur–graphite electrode (Eq. 30).[53] In this process, elemental sulfur is reduced first to form the sulfur dianion, which in turn reacts with unsaturated centers in the heterocycles.

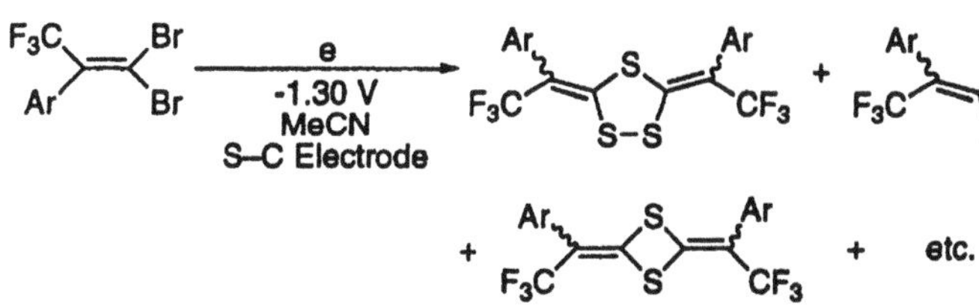

(30)

4. CATHODIC REDUCTION OF OTHER POLYFLUORINATED ORGANIC COMPOUNDS

Polyfluorobenzyl alcohols are a class of fluorinated starting materials which find utility in the areas of medicine and agriculture. Practical electrochemical methods for the selective syntheses of polyfluorobenzyl alcohols have been developed.[54,55] Pentafluorobenzonic acid **6** is reduced selectively to pentafluorobenzyl alcohol **7** at amalgamated lead, zinc and cadmium cathodes in aqueous sulfuric acid solutions, while 2,3,5,6-tetrafluorobenzyl alcohol **8** can also be obtained selec-

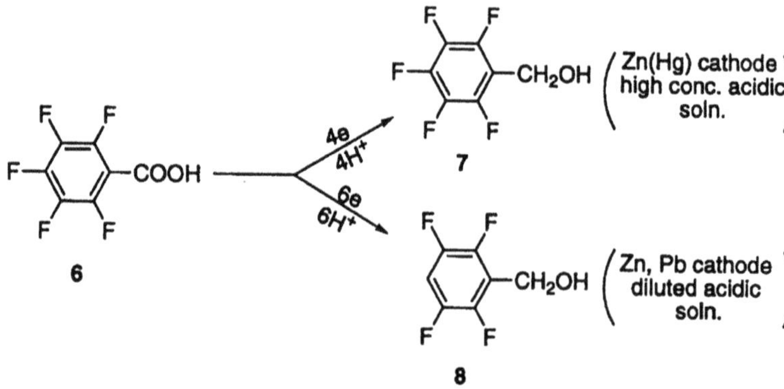

Scheme 10.

tively by using the nonamalgamated cathodes in solutions containing small amounts of a quaternary ammonium salt as shown in Scheme 10.

Interestingly, novel conductive materials have been prepared by cathodic electropolymerization of perfluorocyclobutene and cyclopentene.[56]

5. APPLICATION OF ELECTROGENERATED BASES TO FLUORO-ORGANIC SYNTHESIS

Electrogenerated bases have been shown to be useful for fluoro-organic synthesis. For example, Fuchigami and Nakagawa have found that the electrogenerated 2-pyrrolidone anion deprotonates (trifluoromethyl) malonic ester to form the corresponding stable enolate, which undergoes efficient alkylation with alkyl halides (Eq. 31).[57] This is notable since α-CF_3 enolates are generally unstable with facile defluorination taking place prior to trapping with electrophiles (Eq. 32). In this reaction, the quaternary ammonium countercation of the electrogenerated pyrrolidone anion is essential while ordinary metal countercations such as Na^+ are not effective.

The trifluoromethyl anion is also difficult to generate because it easily loses a fluoride ion to form difluorocarbene. The generation of such a species from trifluoromethane has been similarly achieved using the electrogenerated α-pyrrolidone anion. The efficient tri-

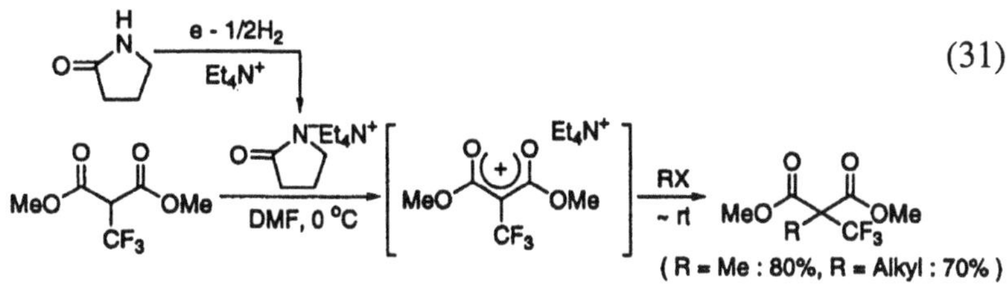

(31)

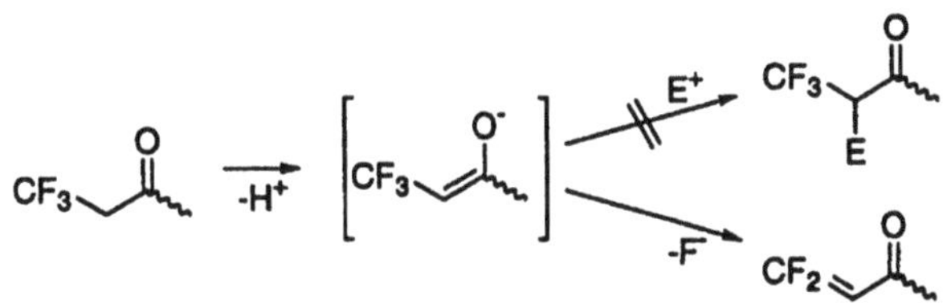

(32)

fluoromethylation of aldehydes and ketones in the presence of hexamethyldisilazane (HMDS) can thus be realized (Eq. 33).[58]

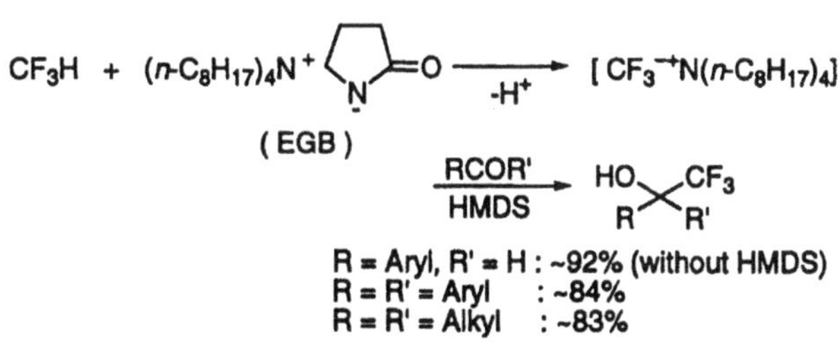

(33)

Quite recently, Troupel has developed an effective synthesis of *gem*-difluoro-β-oxonitriles by using an electrogenerated base derived from bromobenzene and a sacrificial magnesium anode together with a nickel cathode coated with a small deposit of cadmium as shown in Scheme 11.[59]

6. ANODIC OXIDATION OF PERFLUORINATED AND POLYFLUORINATED ORGANIC COMPOUNDS

6.1. Anodic Molecular Conversion

Perfluoroalkyl iodides RfI and polyfluoroalkyl iodides of $RfCH_2CH_2I$ type can be directly oxidized (see the oxidation potential data summarized in Table 4).[60]

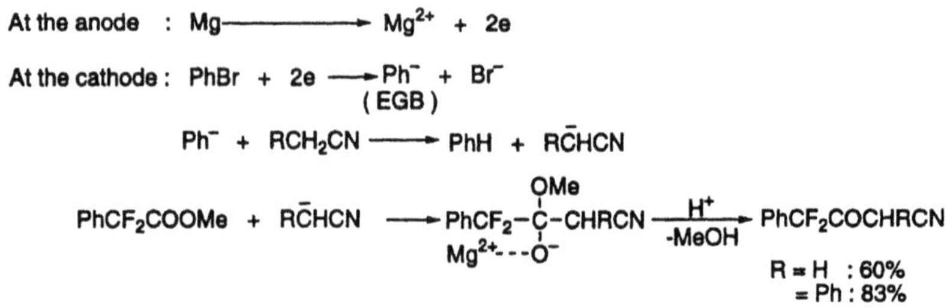

Scheme 11.

Germain and Commeyras have found that perfluorosulfonic esters and fluorosulfates are formed in high yields by direct anodic oxidation of RfI in perfluoroalkane sulfonic acids and fluorosulfuric acid (Eqs. 34 and 35).[61] With diiodo compounds, the mono- and diesters can be selectively obtained (Eq. 36). These are useful precursors to valuable perfluoro carboxylic acids.

$$\mathrm{RfI} \xrightarrow[\mathrm{Rf'SO_3H}]{-2e} \mathrm{Rf'SO_3Rf} + 1/2\ \mathrm{I_2} \qquad (34)$$

$$\mathrm{Rf} = \mathrm{C_2F_5},\ \mathrm{Rf'} = \mathrm{C_4F_9}\ :\ 88\%$$
$$\mathrm{Rf} = \mathrm{C_6F_{13}},\ \mathrm{Rf'} = \mathrm{CF_3}\ :\ 92\%$$

$$\mathrm{C_6F_{13}I} \xrightarrow[\mathrm{FSO_3H}]{-e} \mathrm{FSO_3C_6F_{13}} + 1/2\ \mathrm{I_2} \qquad (35)$$

$$85\%$$

Table 4. Oxidation Potentials (Peak Potentials, E_p) of Perfluoro and Polyfluoro Iodides[a]

Compound	*E_p V vs Ag wire*[b]
C_4F_9I	1.53
$C_6F_{13}I$	1.56
$C_8F_{17}I$	1.56
$C_6F_{13}CH_2CH_2I$	1.55
$C_8F_{17}CH_2CH_2I$	1.39

Notes: [a]In 0.1 M Et_4NBF_4/MeCN; Pt anode; 0.1 V/s.
[b]+1.5 V versus SCE.

$$(C_4F_9SO_3CF_2CF_2)_2 \xleftarrow[C_4F_9SO_3H]{-2e(\text{galvanostatic})} I(CF_2)_4I \xrightarrow[C_4F_9SO_3H]{-e(\text{potentiostatic})} C_4F_9SO_3(CF_2)_4I \quad (36)$$

Since the reaction is shown to be a one-electron oxidation process, it may proceed via a perfluoro carbocation intermediate as is seen with nonfluorinated iodides (Eq. 37). The perfluoro iodide $C_6F_{13}CH_2CH_2I$ is similarly oxidized to give its corresponding sulfonic ester quantitatively (Eq. 38).

$$RfCF_2I \xrightarrow{-e} [RfCF_2I]^{+\bullet} \xrightarrow{-I\bullet} [RfCF_2]^+ \xrightarrow[\text{or } FSO_3^-]{R'fSO_3^-} R'fSO_3CF_2R_f \text{ or } FSO_3CF_2Rf \quad (37)$$

$$C_6F_{13}CH_2CH_2I \xrightarrow[CF_3SO_3H]{-e} CF_3SO_3CH_2CH_2C_6F_{13} \qquad 98\% \quad (38)$$

Becker has also attempted the anodic oxidation of $RfCH_2CH_2I$ in acetonitrile and found that it leads to conversion of $C_8F_{17}CH_2CH_2I$ to the corresponding acetamide, trifluoroacetate, and benzoate derivatives in good yields.[60] A different reaction mechanism is proposed involving a hypervalent iodinyl radical intermediate **C** as shown in Eq. 39.

$$C_8F_{17}CH_2CH_2I \xrightarrow{-e} [C_8F_{17}CH_2CH_2I]^{+\bullet} \xrightarrow{Nu^-} \underset{\mathbf{C}}{[C_8F_{17}CH_2CH_2INu]^\bullet}$$

$$\longrightarrow C_8F_{17}CH_2CH_2Nu + I\bullet\ (1/2\ I_2)$$

$$(Nu^- = MeCN + OH^-, CF_3COO^-, PhCOO^-) \quad (39)$$

Germain has also shown that indirect anodic oxidation in fluorosulfuric acid of fluorocarbon derivatives of type $RfCF_2X$ (X = H, COOH, SO_3H, CH_2OH, Br), which cannot be directly oxidized, leads to fluorosulfates of type FSO_3CF_2Rf (Eqs. 40 and 41).[62,63] In these reactions, the peroxide $(FSO_3)_2$ partially dissociates to form free radicals, which are the in situ electrochemically produced reactive intermediates as shown in Scheme 12.

$$C_3F_7COOH \xrightarrow[FSO_3^-]{-2e} FSO_3C_3F_7 \qquad 85\% \quad (40)$$

$$H(CF_2)_6CH_2OH \xrightarrow[FSO_3^-]{-2e} FSO_3(CF_2)_6O_3SF \quad 82\% \tag{41}$$

$$2FSO_3^- \xrightarrow{-2e} (FSO_3)_2 \rightleftharpoons 2FSO_3\bullet$$

$$FSO_3\bullet + RfH \longrightarrow FSO_3H + Rf\bullet$$

$$FSO_3\bullet + Rf\bullet \longrightarrow FSO_3Rf \qquad Rf = C_2F_5 : 87\%,\ = C_6F_{13} : 92\%$$

Scheme 12.

These products are precursors of perfluoro carboxylic acids while the fluorosulfates derived from secondary and tertiary hydrofluorocarbons provide perfluoro ketones, and starting from the 1*H*-perfluoro bicyclo[2.2.1]heptane, the corresponding perfluoro teriary alcohol is obtained (Eqs. 42 and 43).[64]

$$(Rf)_2CFH \xrightarrow[FSO_3^-]{-2e} FSO_3CF(Rf)_2 \xrightarrow[KF/CsF]{\Delta} (Rf)_2C{=}O \tag{42}$$

$$(Rf)_3CH \xrightarrow[FSO_3^-]{-2e} FSO_3CF(Rf)_3 \longrightarrow (Rf)_3COH \tag{43}$$

6.2. Anodic Polymerization of Polyfluorinated Organic Compounds

Conductive polymers have attracted increasing attention as a result of their wide applications. Recently, very stable poly(thiophenes) with polyfluorinated side chains have been electrochemically synthesized and characterized.[65] Anodic oxidation of 2,3,5,6-tetrafluoroaniline also provides highly stable conductive perfluoropolyaniline.[66]

7. ANODIC OXIDATION OF HETEROATOM COMPOUNDS CONTAINING FLUOROALKYL GROUPS

7.1. General Aspects

As mentioned in the Introduction, partially fluorinated compounds are highly useful. However, methods for their synthesis are strictly

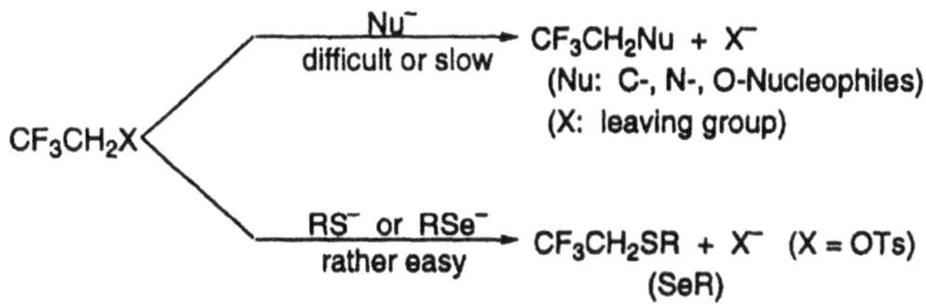

Scheme 13.

limited. For example, nucleophilic substitution occurs with difficulty at the position α to a trifluoromethyl group because of its strong electron withdrawing nature. In contrast, soft nucleophiles such as sulfur and selenium anions undergo substitution reactions at this group rather efficiently (Scheme 13).

Therefore, achievement of a desired substitution reaction, particularly the formation of a carbon–carbon bond at the α-trifluoromethyl position, is an important goal of modern organofluorine chemistry. Although anodic substitution is a characteristic of certain electrolytic reactions, no studies pertaining to the electrolytic substitution of trifluoromethylated compounds have been reported. Recently, the use of the electrochemical technique has opened a new avenue for the realization of such nucleophilic substitution reactions[67–69] and the construction of carbon–carbon bonds.[70–72]

7.2. Oxidation Potentials of Heteroatom Compounds Possessing Fluoroalkyl Groups

Fluoroalkylated Chalcogeno Compounds

The oxidation potentials of various fluoroalkyl sulfides, selenides, and tellurides have been determined by use of cyclic voltammetry.[69,70,73] These substances exhibit irreversible multiple anodic waves. In Table 5 is summarized the first oxidation peak potentials (E_p) of these substances together with those of some nonfluorinated and other electronegatively substituted chalcogeno compounds. In all cases, the fluorinated compounds are oxidized at more positive potentials than the corresponding nonfluorinated compounds because of the electron-withdrawing effects of fluoroalkyl groups. It should be noted

Table 5. Oxidation Potentials (Peak Potentials, E_p) of Fluoroalkyl Chalcogeno Compounds[a]
$PhZCH_2Rf$

	E_p V vs SCE		
Rf	*Z = S*	*Z = Se*	*Z = Te*
CF_3	1.78	1.70	1.34
CF_3CF_2	1.82	1.71	—
$CF_3CF_2CF_2$	1.84	1.72	—
CHF_2	1.69	—	—
CH_2F	1.58	—	—
CH_2CF_3	1.63	1.50	—
H	1.51	1.32	0.82
CN	1.84	1.70	—
COOEt	1.64	1.50	—

Note: [a]In 0.1 M Bu_4NBF_4/MeCN; 0.1 V/s; Pt anode.

that 2,2,2-trifluoroethyl and cyanomethyl selenides show almost equal oxidation potentials although the CF_3 group is expected to have a weaker electron-withdrawing capability than CN based on Taft's σ* values (σ* of CH_2CF_3 = 0.92 vs. 1.30 for CH_2CN). A similar trend is observed in the case of the corresponding amines (see Table 6).[74]

A good linear correlation of the oxidation potentials of sulfides with Taft's σ* values of fluoromethyl groups is obtained as shown in Figure 1.[69] This clearly indicates that the polar effect of the fluoroalkyl group plays a significant role in the electron transfer step from the sulfides to the anode. Namely, the oxidation potential increases linearly as the number of fluorine atoms of the fluoroalkyl group increases. However, interestingly the oxidation potential is not appreciably affected by the length of the perfluoroalkyl group (Table 5).

Fluoroalkylated Amines

Oxidation potentials (half peak potentials, $E_{p1/2}$) of a selection of fluoroethylamines and related nonfluorinated amines are listed in Table 6.[68,74] It is notable that amines with chloromethyl groups such as CF_2Cl and CHFCl show slightly higher oxidation potentials than the corresponding fluoromethyl substituted compounds even though

Table 6. Oxidation Potentials (Half Peak Potentials, $E_{p1/2}$) of Fluoroethylamines and Related Amines

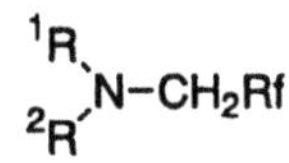

Amine			
R^1	R^2	*Rf*	$E_{p1/2}$ *V vs SCE*
Ph	Me	CF_3	0.96[a]
Ph	Et	CF_3	0.96[a]
Ph	Et	CN	0.95[a]
Ph	Et	Me	0.64[a]
Ph	Et	CF_2Cl	1.00[a]
Ph	Et	CHFCl	0.92[a]
Ph	Et	CHF_2	0.88[a]
Ph	Et	CH_2F	0.76[a]
Bzl	Et	CF_3	1.24[b]
Bzl	Me	H	1.10[b]
n-Bu	*n*-Bu	CF3	1.15[b]
–(CH_2)$_5$–		CF_3	1.18[b]
–(CH_2)$_5$–		H	1.10[b]

Notes: [a]In 0.1 M Et_4NOTs/MeCN; 100 mV/s, glassy carbon anode.
[b]In 0.1 M $NaClO_4$/MeCN; 100 mV/s; Pt anode.

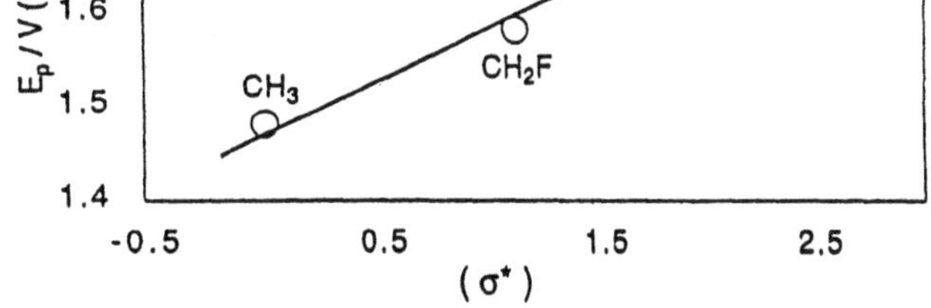

Figure 1. Relationship between oxidation peak potentials (E_p) of fluoroalkyl sulfides and Taft's substtituent constants (σ*) (from Ref. 69).

chorine has a lower electronegativity than fluorine. The oxidation potential of fluoroethylamines increases as the number of fluorine atoms in the fluoroalkyl group increases in a manner similar to that of fluoroalkyl sulfides.

(2-Fluoroalkylallyl)silanes

Allylsilanes are known to be more easily oxidized than the corresponding unsilylated olefins.[75] However, introduction of a fluoroalkyl group to the 2-position of an allylsilane causes a large increase in its oxidation potential as shown in Table 7.[76] This effect is quite different from those observed with fluorinated chalcogeno compounds and amines (Tables 5 and 6).

Table 7. Oxidation Potentials (Peak Potentials, E_p) of (2-Fluoroalkylallyl)trimethylsilanes[a]

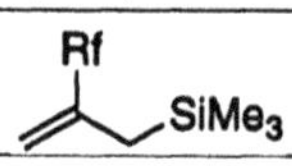

Rf	E_p *V vs SCE*
CF_3	2.63
CHF_2	2.43
COOEt	2.38
H	2.12

Note: [a]0.1 M Bu_4NClO_4/MeCN; Pt anode; scan rate 0.2 V/s.

7.3. Anodic Substitutions of Fluoroalkyl Sulfides

Anodic Methoxylation and Acetoxylation of Fluoroalkyl Sulfides

Anodic substitution reactions of organic nitrogen and sulfur compounds are well-known and they have been used as key reactions for organic synthesis. Recently, Fuchigami and his co-workers have found that aryl 2,2,2-trifluoroethyl sulfides, which are readily derived from inexpensive trifluoroethanol, undergo efficient anodic methoxylation and acetoxylation reactions. In addition, these workers have successfully achieved introduction of oxygen nucleophiles to the position α to the CF_3 group as shown in Scheme 14.[67,70] In contrast, anodic

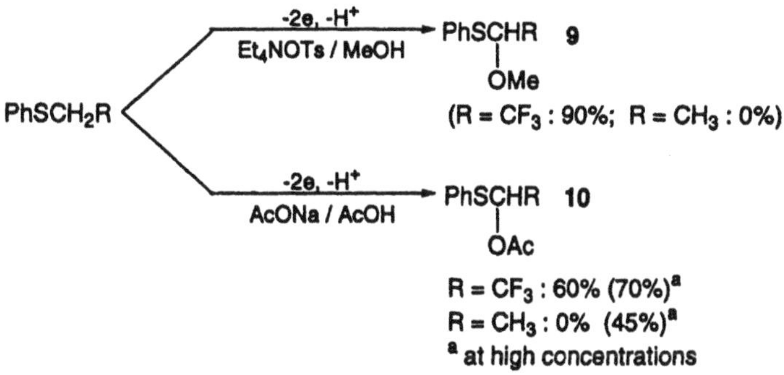

Scheme 14.

methoxylation of nonfluorinated sulfides does not occur. In addition, anodic acetoxylation of simple alkyl sulfides proceeds only when the concentrations of both the substrate and the acetate ions of the supporting electrolyte are extremely high whereas the anodic acetoxylation of 2,2,2-trifluoroethyl sulfides takes place rather efficiently even at low concentrations. Similarly, fluoroalkylated sulfides give much higher yields compared with nonfluorinated sulfides even when the latter substrates are used at high concentrations (Scheme 14).

These facts indicate that an α-trifluoromethyl group markedly promotes anodic substitutions with oxygen nucleophiles. As very few successful examples of anodic methoxylation of phenyl alkyl sulfides are known,[77–79] it is notable that an α-trifluoromethyl group facilitates the anodic methoxylation. As a result, these anodic substitution reactions have been systematically investigated from both a mechanistic and a synthetic viewpoint.[69]

It was found that anodic methoxylation of aryl 2,2,2-trifluoroethyl sulfides is greatly affected by substitution on the aryl ring. Electron-withdrawing groups promote anodic methoxylation while electron-donating groups significantly retard this process (Table 8). As shown in Table 8, the product yields are better correlated to Hammett's σ^+ values than to σ values, namely, as the σ^+ value becomes more negative, the yield decreases. Therefore, the reaction is not governed by the stability of the cationic intermediates **D** or **E** as illustrated in Scheme 15. Because the electron-withdrawing substituent (X = Cl) promotes the methoxylation as observed in the case of nonsubstituted

Table 8. Efficiencies of the Anodic Methoxylation and Hammett's σ and σ^+ Values of Substituents (X)

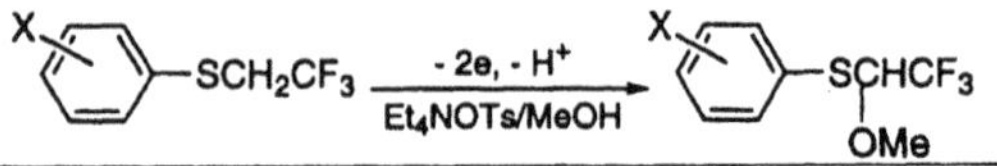

X	p-Cl	H	p-F	m-Me	p-Me	p-MeO
Yield (%)	80	93	43	28	18	13
σ^+	+0.11	0	−0.07	−0.07	−0.31	−0.78
σ	+0.23	0	+0.66	−0.07	−0.17	−0.27

phenyl sulfide, this reaction must be controlled by the ease of deprotonation of the radical cation intermediate **D** (step a).

Anodic methoxylation and acetoxylation are also significantly affected by the structure of the fluoroalkyl group (see Table 9). Strong electron-withdrawing perfluoroalkyl groups promote the anodic methoxylation. In contrast, a difluoromethyl group causes a drastic decrease in the yield of the process and monofluoromethyl and chlorofluoromethyl groups do not promote the methoxylation at all.

In contrast to the anodic methoxylation, the anodic acetoxylation takes place regardless of the fluoroalkyl group and the yield enhancement effect is in the order $CF_3 > C_2F_5 > C_3F_7 >> CHF_2 > CH_2F >$ CHClF. This order, which is similar to that observed for the methoxylation reaction, can be rationalized as being related to the ease of deprotonation of **D** (i.e., kinetic acidity of **D**). Thus, the anodic methoxylation and acetoxylation reactions of aryl fluoroalkyl sulfides are greatly affected by both substituents on the benzene ring and fluoroalkyl groups.

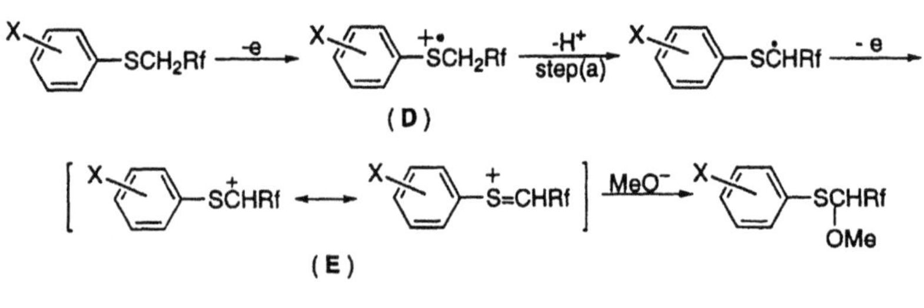

Scheme 15.

Table 9. Anodic Methoxylation and Acetoxylation of Fluoroalkyl Phenyl Sulfides

$$PhSCH_2Rf \xrightarrow[YO^-]{-2e,\ -H^+} PhSCH(OY)Rf$$

Rf			*Yield (%)*			
Y	*CF_3CF_2*	*CF_3*	*CHF_2*	*CH_2F*	*CHClF*	*CH_3*
Me	72	93	19	—[b]	—[c]	0
Ac[a]	46	60	28	20	16	0

Notes: [a]At low concentrations.
[b]Many complicated products were formed.
[c]Not detected.

Trifluoroacetaldehyde is difficult to prepare from trifluoroethanol. α-Methoxylated and α-acetoxylated 2,2,2-trifluoroethyl sulfides are trifluoroacetaldehyde equivalents, which are useful building blocks for synthesis of various trifluoromethylated compounds. The α-acetoxy sulfide **10** is easily converted into trifluoroacetaldehyde phenylhydrazone (Eq. 44). The α-methoxy sulfide **9** is also readily transformed into biologically interesting α-monofluoroalkanoic acids (Eq. 45).

$$\underset{\mathbf{10}}{CF_3CH(OAc)SPh} \xrightarrow{aq.K_2CO_3} [CF_3CHO] \xrightarrow{PhNHNH_2} CF_3CH{=}NNHPh \quad 70\% \tag{44}$$

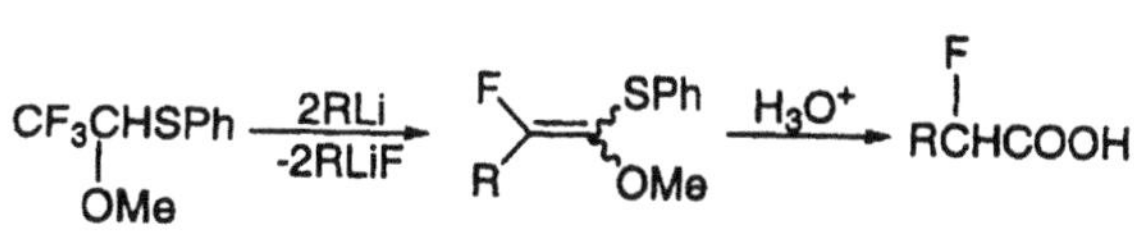

9

R = *n*-Bu : 95%
s-Bu : 80%
Ph : 31

(45)

Because these methoxylated and acetoxylated sulfides **9**, **10** have an acetal structure, it is expected that Lewis acid-catalyzed demethoxylation should generate a carbocation intermediate which is stabilized by the neighboring sulfur. In fact, nucleophilic substitution with arenes has been successfully achieved as shown in Eq. 46.[70] This methodology is useful for the preparation of trifluoroethyl arenes, reductive desulfurization being readily performed as shown in Eq. 47. As already mentioned, generation of carbocations bearing an α-trifluoromethyl group is difficult because of the strong electron-withdrawing effect of this substituent. Therefore, this carbon–carbon bond formation reaction is notable from both a mechanistic and a synthetic perspective.

$$\underset{\mathbf{9}}{CF_3CH(OMe)SPh} \xrightarrow[-\ MeO^-]{AlCl_3} [CF_3\overset{+}{C}HSPh \rightleftharpoons CF_3CH{=}\overset{+}{S}Ph] \xrightarrow[-H^+]{ArH} CF_3CH(Ar)SPh$$

Ar = *p*-(*i*-Bu)C_6H_4 (48%); Ph (83%)

(46)

$$CF_3CH(Ar)SPh \xrightarrow[-Bu_3SnSPh]{Bu_3SnH} \underset{\text{quant.}}{CF_3CH_2Ar} \tag{47}$$

Nucleophilic substitution at the β-position to a trifluoromethyl group is also generally difficult, except for sulfur nucleophiles owing to predominant elimination to form trifluoropropene as shown in Eq. 48.

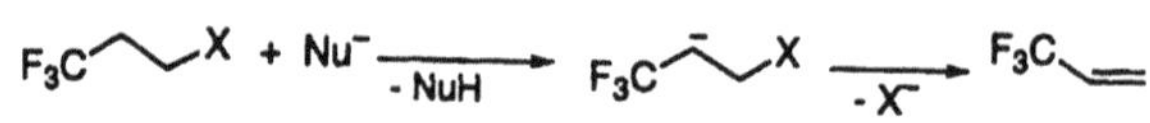

(48)

Strongly basic carbon, oxygen, and nitrogen nucleophiles initiate E2 eliminations by deprotonation to give the corresponding olefins. In the case of E1 reactions, facile deprotonation from the β-position of the cationic intermediate takes place. The nucleophile may then be added to the double bond. It has been demonstrated that such nucleophilic substitution can be performed by anodic oxidation of polyfluoroalkyl iodides ($RfCH_2CH_2I$) in the presence of several oxygen and nitrogen nucleophiles (see Eq. 39). β-Trifluoromethylated *O,S*-acetals seem to be promising building blocks for this purpose. The anodic acetoxylation of phenyl 3,3,3-trifluoropropyl sulfide at high concentrations provides the corresponding α-acetoxy sulfide **11** in satisfactory yield (Eq. 49) while a conventional Pummerer reaction approach gave much lower yields.[80] Therefore, the electrochemical method is superior to the conventional Pummerer reaction as the α-acetoxl sulfide can be obtained in one step under mild conditions.

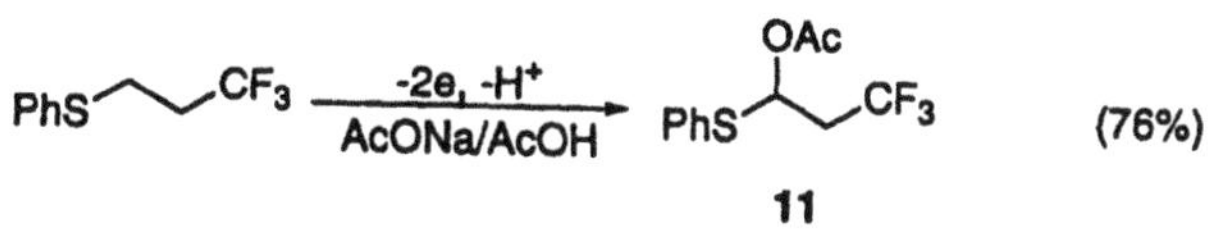

(49)

The thus prepared β-trifluoromethylated *O,S*-acetal **11** is a highly versatile building block for making carbon–carbon bonds via a carbocation at the β-position to the CF_3 group as shown in Scheme 16.[80] Notably, Lewis acid-mediated allylation and cyanation can be achieved efficiently when electrogenerated acids are employed.

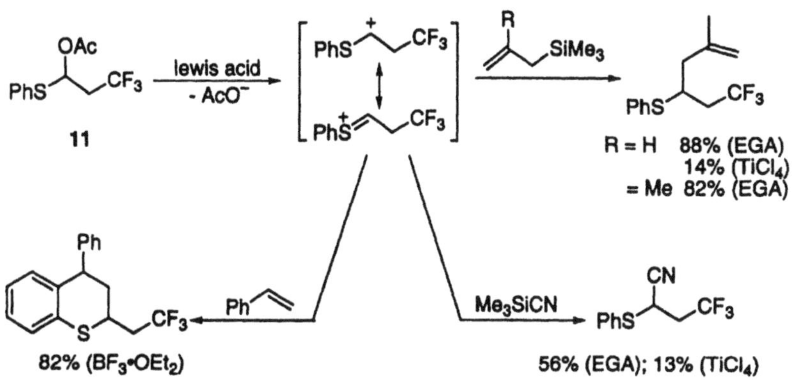

Scheme 16.

Fluoride Ion-Mediated Anodic Methoxylation

As explained in detail above, Fuchigami and his co-workers observed that the CF_3 group promotes anodic fluorination reactions of sulfides.[81] Interestingly, the CHF_2 and CH_2F groups also promote these processes (Eq. 50).[82] This stands in sharp contrast to earlier reports indicating that the CH_2F group does not promote anodic methoxylation of sulfides (see above). Interestingly, anodic oxidation of 2-monofluoroethyl sulfide in methanolic $Et_3N{\cdot}3HF$ instead of acetonitrile as a solvent provides the α-methoxylated product **13** exclusively. Also, the α-fluorinated product **12** is not formed in this process (Eq. 51).[82,83] This novel fluoride ion-promoted anodic methoxylation is notable because anodic methoxylation of sulfides is normally very inefficient unless a strong electron-withdrawing group[68,79] or an excellent leaving group (e.g., silyl)[77,78] is present.

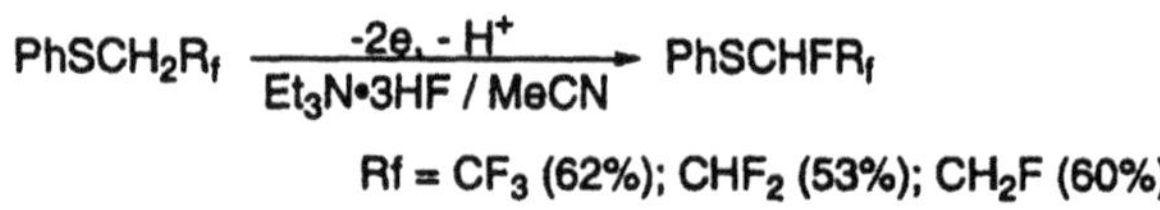

(50)

(51)

Among the fluorides, $Et_3N{\cdot}3HF$ is the most effective in promoting this anodic oxidation reaction and halide ions other than fluoride ion are not effective at all. It is notable that $Bu_4NF{\cdot}3H_2O$ is also effective and does not cause formation of a sulfoxide through water capture of the initially formed, sulfur-centered cation radical.

The fluoride ion promoted anodic substitution process is widely applicable to a number of systems (see Table 10). The current efficien-

Table 10. Fluoride Ion Promoted Anodic Methoxylation of Fluoroalkyl Sulfides

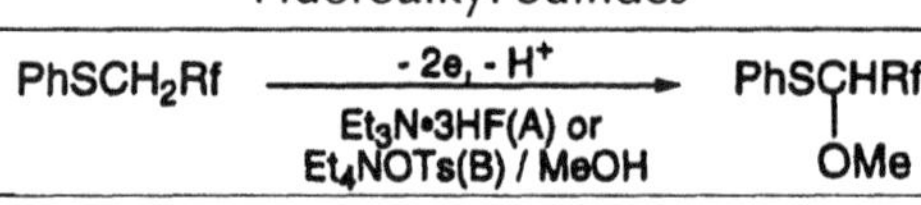

Sulfide Rf	*Supporting Electrolyte*	*Charge Passed*	*Product Yield (%)*
CH_2F	A	3.5	63
CH_2F	B	3.5	0
CHF_2	A	5.1	98
CHF_2	B	10	19
CH_2CF_3	A	3.5	74
CH_2CF_3	B	10	trace

cies and the yields of the desired methoxylated products are greatly improved by use of this methodology. Further demonstrating the utility of this process is the fact that $PhSCH(OMe)CH_2CF_3$ is a useful synthetic building block, serving as a CF_3CH_2CHO equivalent.

This novel anodic methoxylation may proceed via the intermediacy of fluorosulfonium ion **G** by way of a Pummerer-type mechanism as shown in Scheme 17. In this route, the sulfide cation radical **F** is trapped by fluoride ion, a process which suppresses potentially complicating side reactions (such as dimerization and nucleophilic attack on an aromatic ring) of this intermediate. Because fluoride ion is a much weaker nucleophile than methoxide, it is reasonable that methoxylation predominates in methanol. Thus, fluoride ions are not incorporated into the products yet they promote the anodic methoxylation, playing the role of a mediator.

In contrast to this fluoride ion mediator effect, it has been established that bromosulfonium ions derived from sulfides and anodically generated bromonium ions can be used as so-called mediators for indirect anodic oxidations of alcohol.[84] Therefore, it is interesting that the fluorosulfonium ions (**G**) also noticeably promote the anodic methoxylation.

This novel fluoride ion mediated methoxylation is widely applicable as shown in Eq. 52.[85–89] Anodic oxidations of 2-ethoxycarbonyl-3,3,3-trifluoropropyl sulfide (**14**), 2-methoxy- and 2-hydroxy-3,3,3-trifluoropropyl sulfides (**15**) and (**16**) were carried out in methanol to give the corresponding α-methoxylated products **17–19** in good yields. Notably, reaction of **14** provides α-(trifluoromethyl)acrylate **20** selectively when acetonitrile is used as solvent instead of methanol.

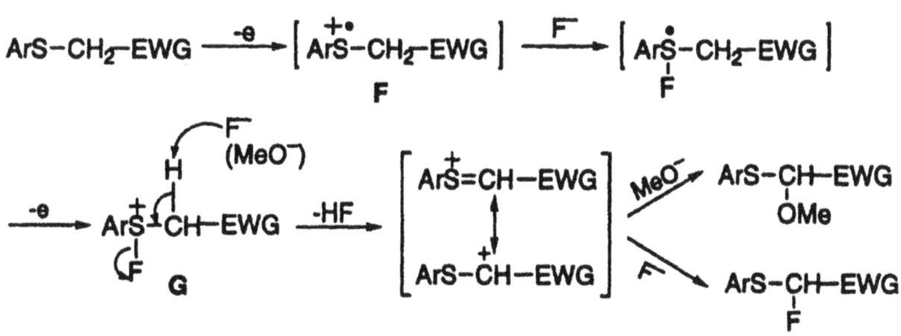

Scheme 17.

In both cases, nonfluoride-containing halide salts were not effective in promoting these processes.[85] These reactions most likely proceed via fluorosulfonium ions as a key intermediate as shown in Scheme 18. In the case of **14**, deprotonation seems to proceed smoothly owing to the strongly electron-withdrawing CF_3 and EtO_2C groups to form **20** selectively.

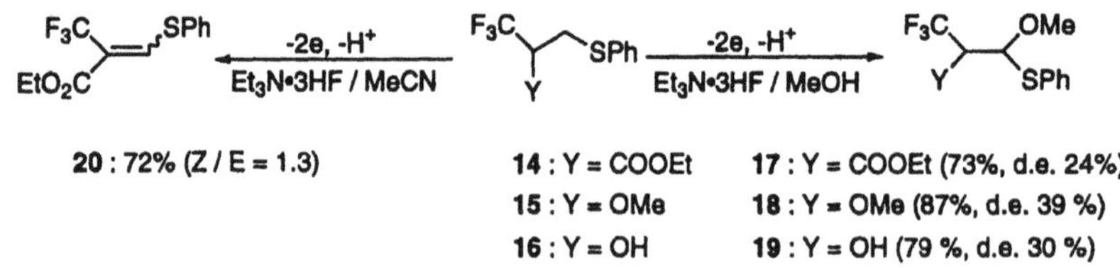

(52)

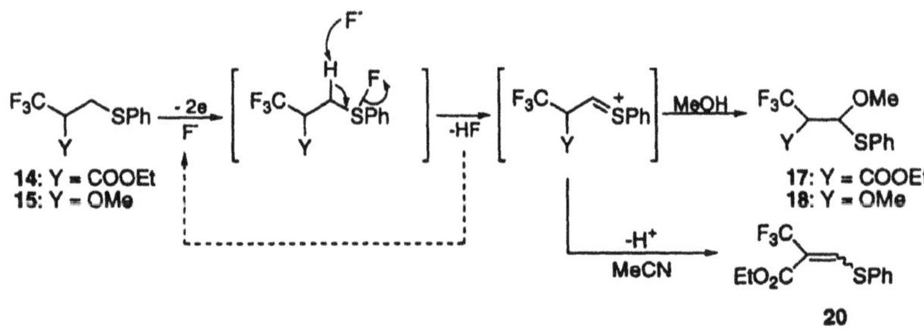

Scheme 18.

The α-methoxylated products obtained in this manner are highly useful building blocks for making a carbon–carbon bond β to the CF_3 group as shown by the examples in Eq. 53.[85,86] It is notable that $TiCl_4$-mediated carbon–carbon bond formation is a highly diastereoselective process because the reaction proceeds via a titanium-chelated intermediate as shown in Eq. 54.[87] Furthermore, anodic

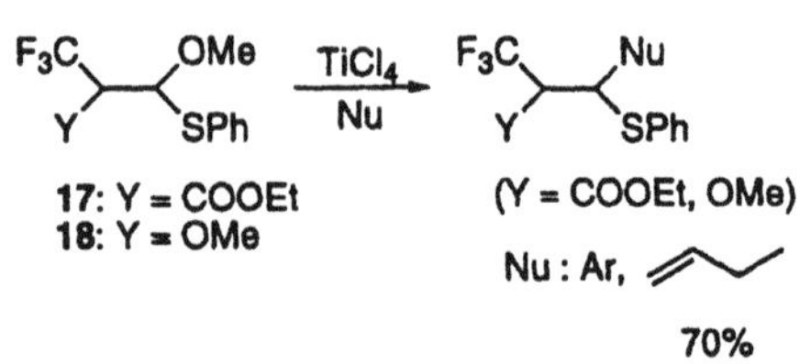

(53)

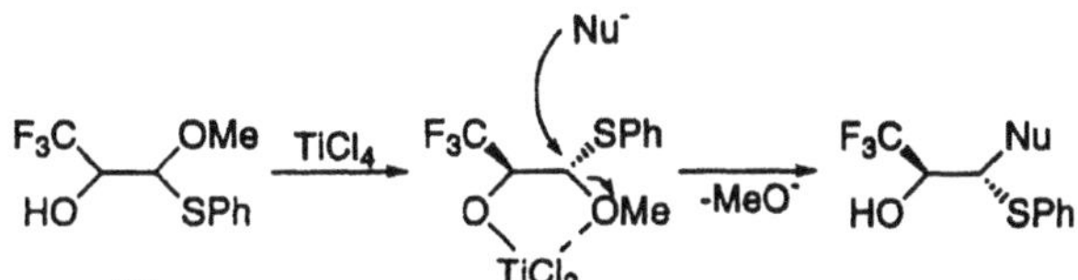

19

(d.e. 30%)

Nu : Ph (72%, d.e. 54%)
p-MeC_6H_4 (65%, d.e. 79%)
(63%, d.e. 86%)

(54)

intramolecular alkoxylation has been successfully performed for the first time by using a fluoride ion mediator as shown in Eq. 55.[88]

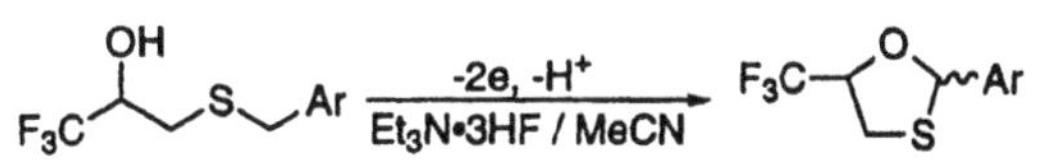

(55)

Ar = Ph (73%, t/c = 1)
p-ClC_6H_4 (64%, t/c = 1.3)
o-ClC_6H_4 (51%, t/c = 1.9)

7.4. Anodic Methoxylation and Acetoxylation of Fluoroalkyl Selenides

The fluoride ion-mediated methoxylation process described above also can be applied to fluoroalkyl selenides. Anodic α-substitution of organic selenium compounds to our knowledge has not been described except in one report by Jouikov.[89] As shown in Eq. 56, even perfluoroalkyl groups do not promote the anodic methoxylation of selenides under conventional conditions using Et_4NOTs as a supporting electrolyte. This is quite different from the results observed in the case of the corresponding sulfides (see Table 9). However, in the presence of fluoride ions, anodic methoxylation proceeds smoothly to give α-methoxylated selenides in relatively good yields (Eq. 56).[90,91]

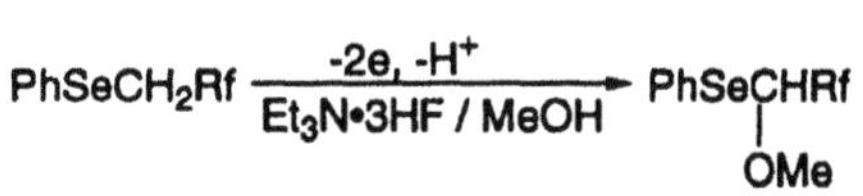

(56)

Rf = C_nF_{2n+1} (n = 1-3) : 65-74%
Et_4NOTs / MeOH : 3-15%

Also, anodic acetoxylation of selenides is markedly promoted by a perfluoroalkyl group (Eq. 57).[73] It is notable that the effect of a CF_3

group on this process is more pronounced than that of a CN group although the CF_3 group has a weaker electron-withdrawing ability.

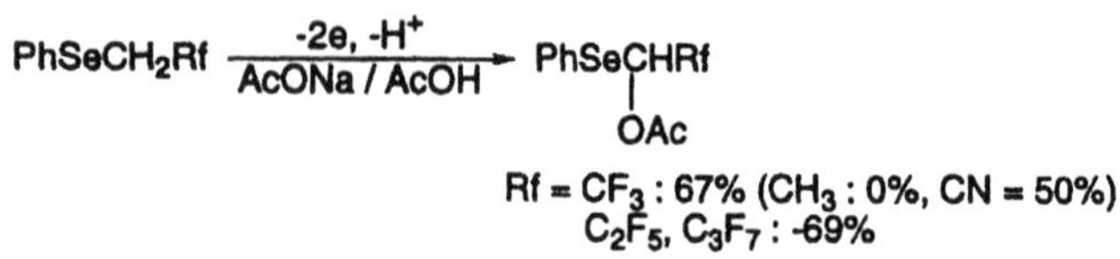

(57)

These methoxylated and acetoxylated selenides are α-perfluoroalkyl monoselenoacetals, which could serve as useful building blocks in a manner similar to that of their sulfur analogues. Thus far, only a limited number of methods have been developed for the preparation of monoselenoacetals and these require rather complicated procedures or specialized reagents. In this regard, the electrochemical method is advantageous as the monoselenoacetals can be prepared in a one-step reaction under mild conditions.

7.5. Anodic Oxidation of Fluoroalkyl Tellurides

In contrast to the cathodic reduction of organic tellurium compounds, few studies on their anodic oxidation have been performed. There are no reports dealing with electrolytic reactions of fluorinated tellurides, which probably reflects the difficulty in preparing the partially fluorinated telluride starting material. Recently, Fuchigami and his co-workers have investigated the anodic behavior of 2,2,2-trifluoroethyl phenyl telluride in an Et_4NOTs MeOH solvent system. Unexpectedly, both methoxylation and *p*-tosyloxylation at tellurium take place selectively to yield hypervalent compounds having Te–O bonds (Eq. 58).[92] In this case, α-methoxylation does not occur, a different outcome than from anodic oxidations of the corresponding sulfides and selenides. Similarly, anodic acetoxylation and benzoxylation resulted at the tellurium atom to provide monomeric and anhydrous dimeric hypervalent products, respectively (Eq. 59).[93]

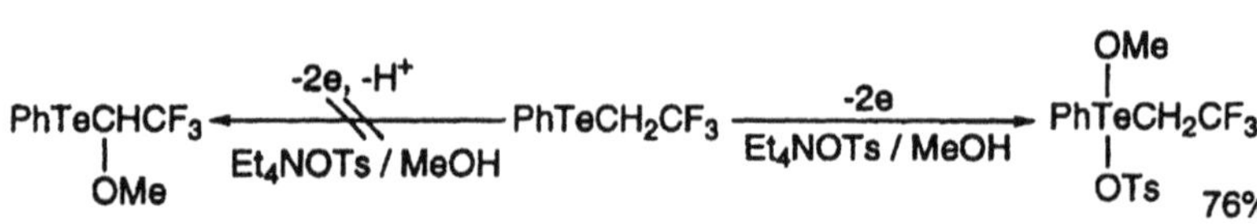

(58)

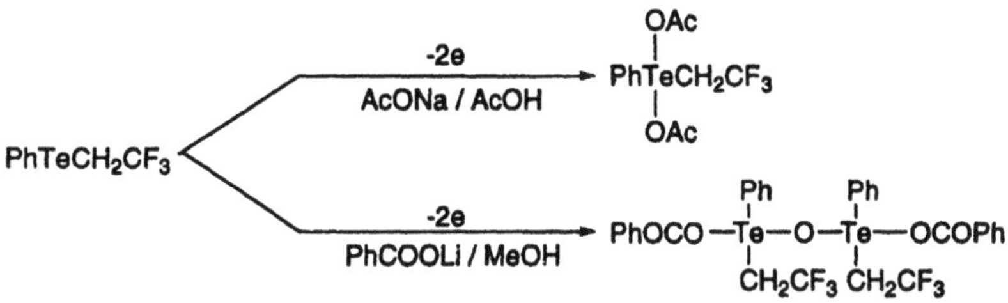

(59)

p-Toluenesulfonate ion, being a poor nucleophile, is commonly used as the counteranion of supporting electrolytes. However, this ion acted as an efficient nucleophile in this process. This notable observation is reasonable, however, because electronegative ligands are generally better stabilizers of hypervalent compounds.

7.6. Anodic Substitutions of Fluoroalkyl Amines

Fuchigami and his co-workers have systematically studied the anodic α-methoxylation reactions of various types of *N*-(fluoroethyl) amines, $ArN(R)CH_2Rf$ (e.g., Rf = CF_3, CHF_2, CH_2F), at a graphite anode in alkaline methanolic solutions by using an undivided cell.[68,74,94] A methoxy group is exclusively or preferentially introduced into a position α to the fluoromethyl (Rf) group depending on the nature of the Rf and R groups (Table 11). *N*-Ethyl derivatives having a CF_3 group give only one regioisomer of the α-methoxylated product (runs 2 and 3 in Table 11).

It is quite interesting that anodic methoxylation also occurs predominantly (71%) α to the 2,2,2-trifluoroethyl group of a substituted *N*-methylaniline (run 1) and in a minor way (29%) at the methyl group leading to formation of $PhNHCH_2CF_3$. Such anodic methoxylations are known to occur exclusively at the methyl group of nonfluorinated *N*-ethyl-*N*-methylaniline.[95] Thus, the CF_3 group dramatically changes the regioselectivity of such anodic methoxylations as shown in Figure 2. The CHF_2 group as well as the trifluoromethyl group markedly promote anodic α-substitution in the case of the *N*-methyl derivatives, where methoxylation at the methyl group is competitive. Notably, even the monofluoromethyl group promotes anodic α-methoxylation (run 9) although the yield in this case is low. Consequently, the order

Table 11. Anodic Methoxylation of Fluoroethylamines

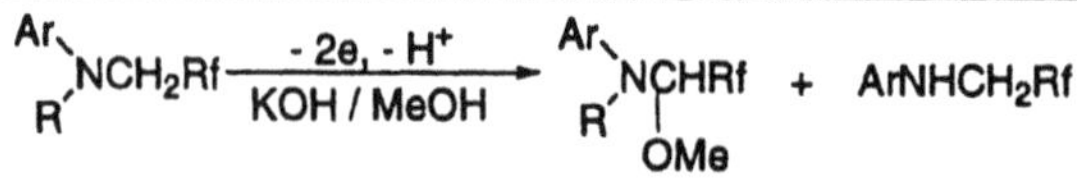

21 22

Run	*Amine*			*Charge Passed (F/mol)*	*Yield (%)*	
	Ar	*R*	*Rf*		**21**	**22**
1	Ph	Me	CF_3	3.8	71	29
2	Ph	Et	CF_3	4.9	85	0
3	*p*-Tol	Et	CF_3	3.5	78	0
4	Ph	Ph	CF_3	7.8	81	0
5	Ph	Me	CHF_2	3.3	19	48
6	Ph	Et	CHF_2	4.5	91	0
7	*p*-Tol	Et	CHF_2	3.5	87	0
8	Ph	Me	CH_2F	2.8	0	—[a]
9	Ph	Et	CH_2F	3.0	45	—

Note: [a]$PhN(CH_2OMe)CH_2CH_2F$ (84%) was formed.

of the amine anodic oxidation promotion effect appears to be CF_3 > CHF_2 > > CH_2F.[74]

The reaction proceeds via electrogenerated cationic intermediates as is seen in the anodic oxidation of nonfluorinated amines, carbamates, and amides (Scheme 19). However, the regiochemistry in these processes is not governed by the stability of the cationic intermediates **I** and **I′** because the main products are formed via the less stable intermediates **I**. Indeed, the promotion effect and unique regioselectivity observed in reactions of the fluoroalkyl amines can be explained mainly in terms of the α-CH kinetic acidities of the cation

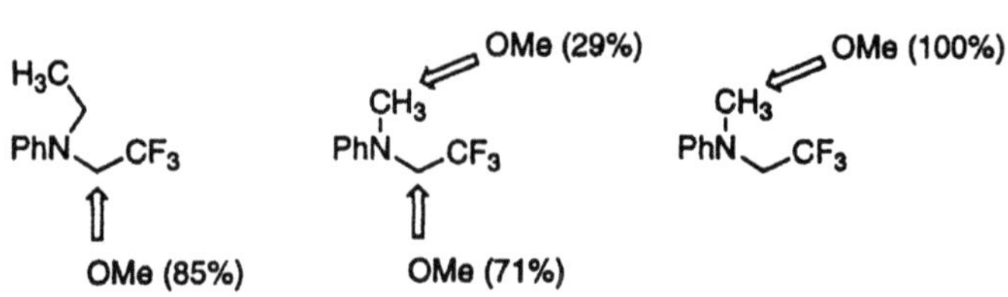

Figure 2.

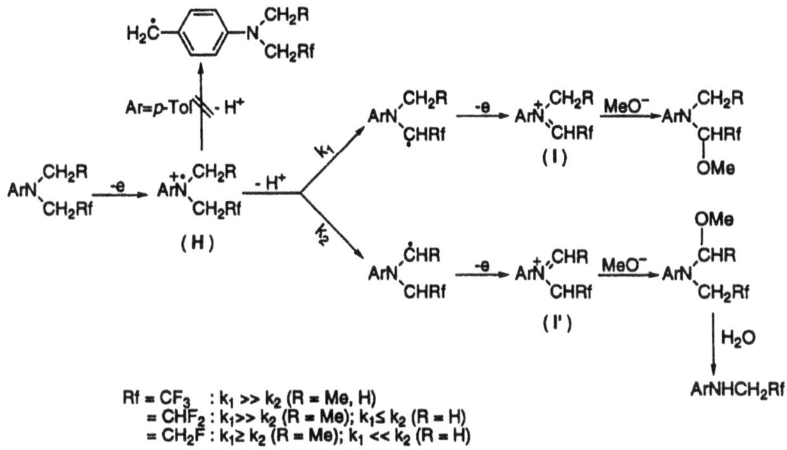

Scheme 19.

radicals **H** formed initially by one-electron oxidation of the amines. Accordingly, the rates of deprotonation are governed by the stronger thermodynamic acidities of the α-methylene hydrogens.

The α-methoxylated products are highly useful synthetic building blocks for making carbon–carbon bonds α to trifluoromethyl and difluoromethyl groups, as shown in Scheme 20. Thus, α-tri- and α-difluoromethylated α-aminonitriles, which are precursors to the corresponding fluorinated α-amino acids, have been prepared in this manner in excellent yields. Tri- and difluoromethylated tetra- and dihydroquinoline derivatives can also be accessed by cationic polar cycloaddition in high yields.

Regioselective anodic methoxylation of more complicated *N*-(tetrafluoropropyl)amines **23** is also successful as shown in Eq. 60. Methoxylation takes place at the fluoroalkyl group preferentially.[96]

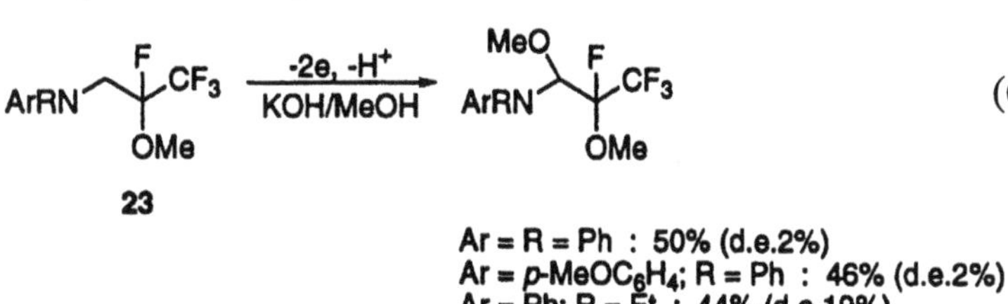

(60)

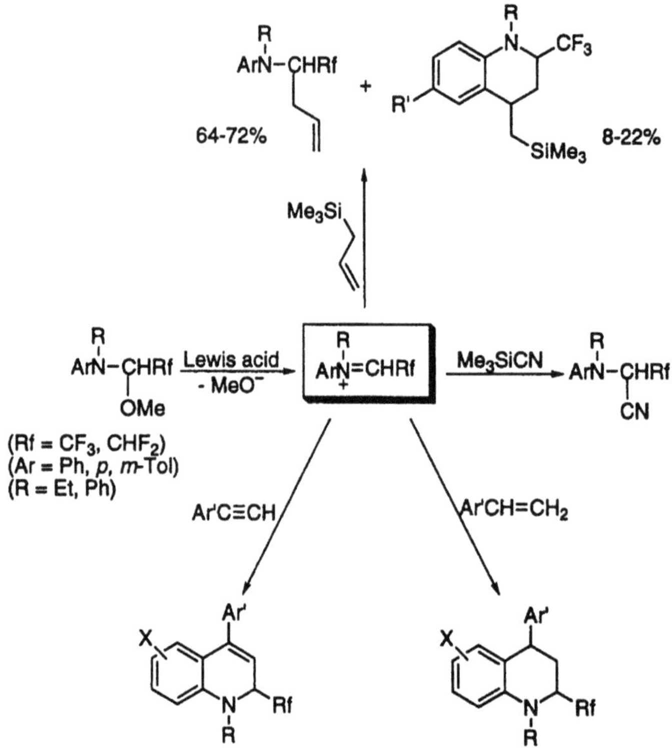

Scheme 20.

It has sometimes been proposed that unique regioselectivities observed in anodic substitution reactions are caused by the heterogeneous nature of electrode reactions. However, Konno and Fuchigami have recently demonstrated that photoaddition of *N*-(2,2,2-trifluoroethyl)-*p*-toluidine to 3-phenylcyclohex-2-en-1-one occurs at the position α to the CF_3 group predominantly.[97] Interestingly, the regioselectivity of this photoaddition is comparable to that of anodic methoxylation (Scheme 21 vs. Fig. 2). Because this photochemical process is initiated by a homogeneous single electron transfer step, the results show that the unique regiochemistry observed in the anodic methoxylation of *N*-(2,2,2-trifluoroethyl)amines is not related to the heterogeniety of the system but rather to control offered by amine cation radical kinetic acidity.

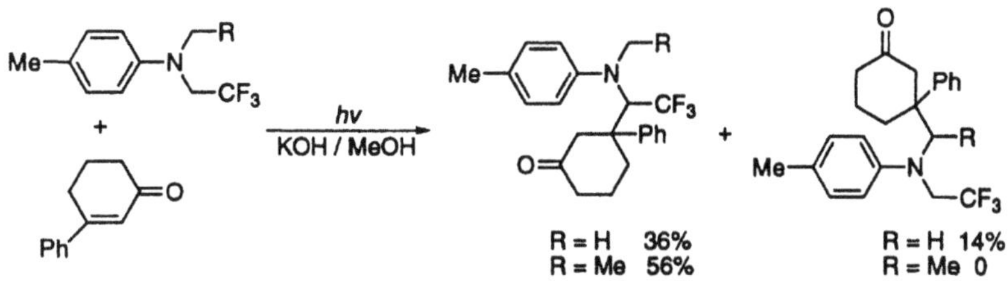

Scheme 21.

Anodic cyanation of amines is a promising method for the preparation of α-amino nitriles, themselves versatile synthetic intermediates. In sharp contrast to the anodic methoxylation of *N*-(2,2,2-trifluoroethyl)amines, anodic cyanation does not occur at the position α to the CF_3 group (Eq. 61).[98] Notably, amine **24**, which has no α-protons other than those in the trifluoroethyl group, does not undergo anodic α-cyanation (Eq. 62). These observations indicate that the CF_3 group completely inhibits cyanation at an α-position, regardless of the molecular structure of the amines. Although it is not clear why the regioselectivity of the anodic cyanation differs from that seen in the related anodic methoxylation processes, differences in the reaction medium and/or in the basicity and nucleophilicity of CN and MeO anions must play a significant role in governing anodic oxidation regioselectivity.

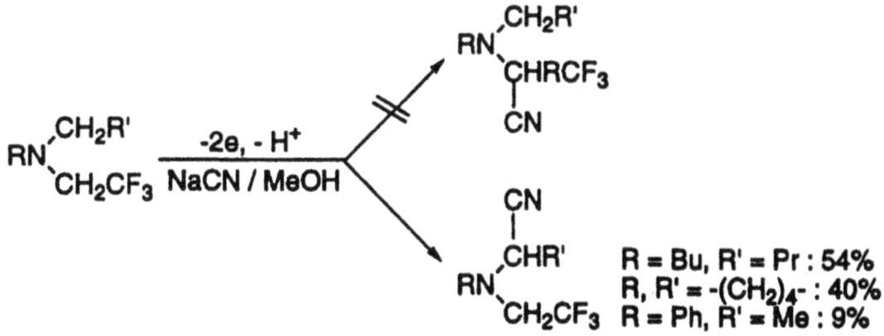

(61)

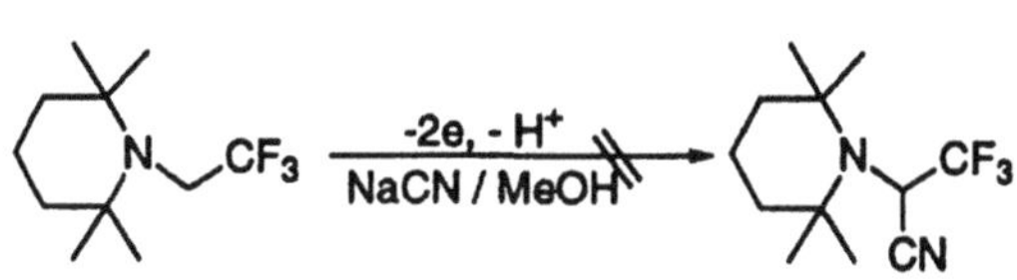

24

(62)

In contrast to anodic methoxylation of 2,2,2-trifluoroethylamines, *N*-(2,2,2-trifluoroethyl)carbamates do not undergo this oxidation reaction in methanol. However, the use of trifluoroethanol as the solvent or cosolvent with acetonitrile for this reaction leads to formation of α-methoxylated product **25** in high yield as shown in Eq. 63.[99] The product **25** is useful as an α-aminotrifluoroethylating reagent in reactions with ketones and malonate esters.

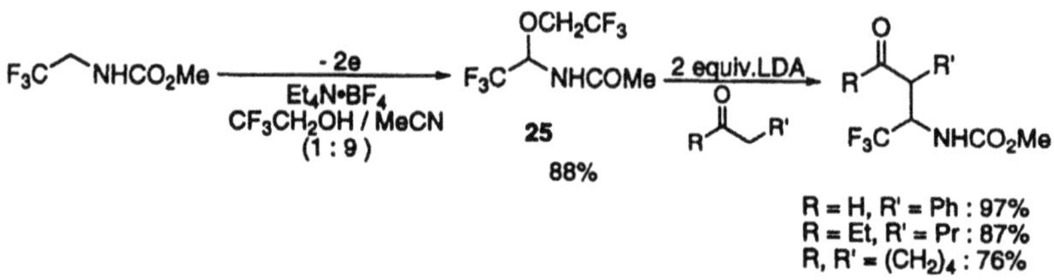

(63)

7.7. Anodic Cyclization of Fluoroalkylamino Compounds

Anodic oxidation of *N,N'*-disubstituted trifluoroethanimidamines **26** in dry acetonitrile gives *N*-substituted 2-(trifluoromethyl)benzimidazoles **27** as shown in Scheme 22.[100,101] Electron-donating *para*-substituents in the *N,N'*-diaryl derivatives markedly promote this cyclization process, whereas *N*-alkyl derivatives react to form *p*-benzoquinone imine derivatives **28** under the anodic oxidation conditions.

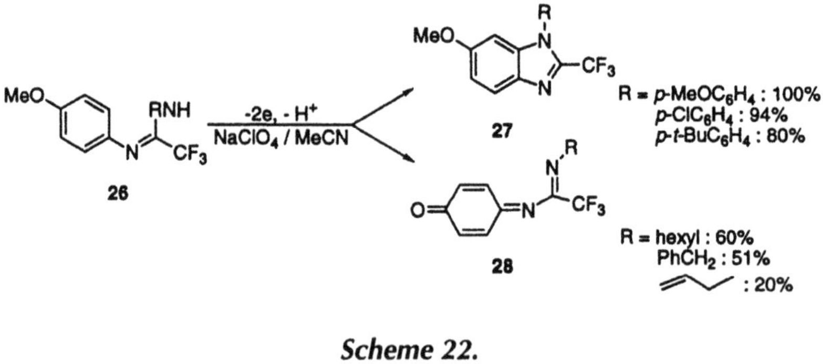

Scheme 22.

Trifluoromethyl-indoles can be prepared in a similar manner as shown in Eq. 64.[102] In this case, anodic oxidation provides *p*-benzoquinone imine derivatives **29**, which are easily converted into trifluoromethyl-indoles **30** by heating or by treatment with ceric ammonium nitrate.

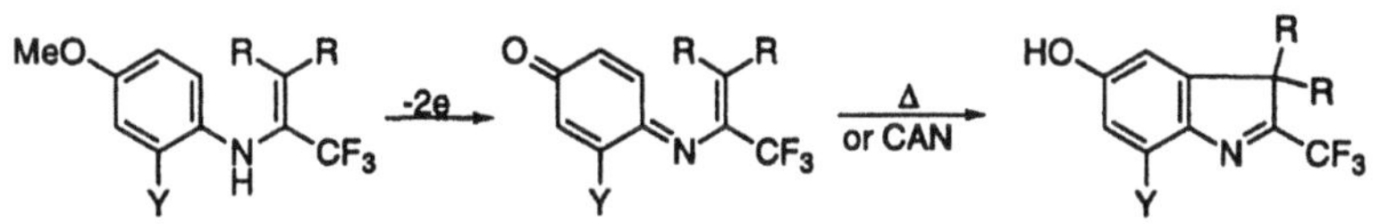

29 **30**

Y = H, R = COOMe : 91%
Y = H, R = CN : 94%
Y = OMe, R = COOMe : 90%

Y = H, R = COOMe : 66%
Y = OMe, R = COOMe : 63%

(64)

7.8. Electrochemical Synthesis of Fluoroalkylated Sulfenimines

Sulfenimines are versatile building blocks for the preparation of secondary and tertiary amines. Therefore, fluoroalkylated sulfenimines should be promising starting materials for the preparation of *N*-fluoroalkylamino compounds. In fact, trifluoromethylated sulfenimines have been prepared recently in one step by indirect anodic oxidation of 2,2,2-trifluoroethylamine and diaryl disulfides in $MeCN/Et_4NClO_4$ by using $MgBr_2$ as a redox mediator (Eq. 65).[103] Both anodically formed "Br^+" and the cathodically generated base $Mg(OH)_2$ are necessary for this reaction.

$$CF_3CH_2NH_2 + ArSSAr \xrightarrow[MgBr_2\,/\,MeCN]{-2e,\ -H^+} [CF_3CH_2NHSAr] \xrightarrow{-2e,\ -2H^+} CF_3CH=NSAr \quad (65)$$

Ar = Ph : 72%
= *p*-Tol : 58%

Sulfenimines have been shown to be highly useful building blocks for the preparation of trifluoromethylated amines, aminoketones, and aminoalkanoates as illustrated in Scheme 23.

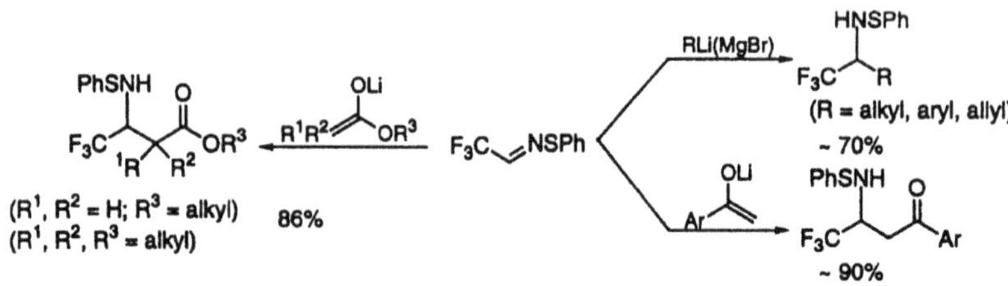

Scheme 23.

Similar indirect anodic oxidations of various fluoropropylamines and diphenyl disulfide in two-phase systems such as CH_2Cl_2–water provide useful fluorinated sulfenimines as shown in Eq. 66.[104]

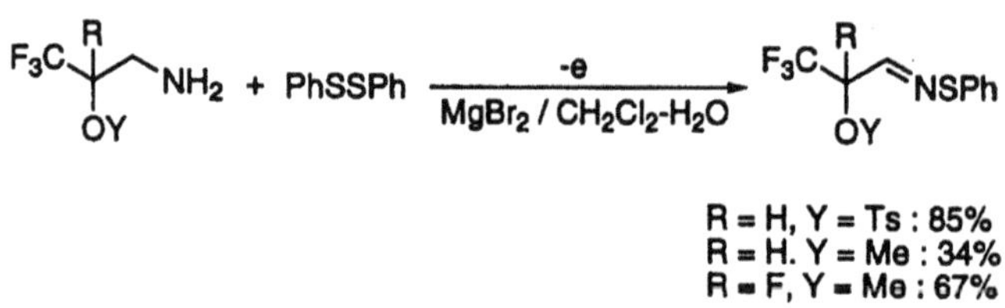

(66)

8. ANODIC OXIDATION OF TRIFLUOROMETHYLATED CARBOXYLIC ACIDS AND TRIFLUOROMETHYLSULFINIC ACID

8.1. Anodic Oxidation of Trifluoroacetic Acid

Electrooxidative generation of the trifluoromethyl radical ($CF_3^{\bullet}$) and its synthetic application has been thoroughly developed since the early 1970s, reflecting the fact that trifluoroacetic acid (TFA) is one of the most readily available and economical starting materials for trifluoromethylation reactions.[105] Heteroaromatics as well as olefins have been employed as trapping agents for this radical leading to overall trifluoromethylation of these substances (Eq. 67).[106]

However, the selectivity for the trifluoromethylation reactions has been rather low in many cases. For example, anodic trifluoromethylation of olefins provides a mixture of several products in general as shown in Eq. 68 and control of the product selectivities has been

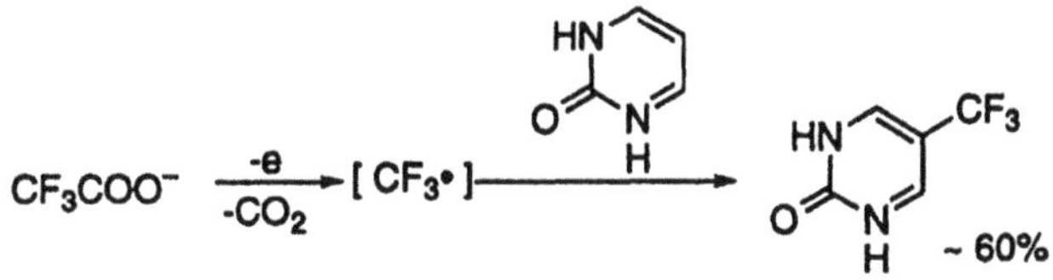

(67)

difficult.[107] Muller has found that the unwanted formation of the dimers can be suppressed by carrying out the reactions in aqueous acetone or methanol.[108–112]

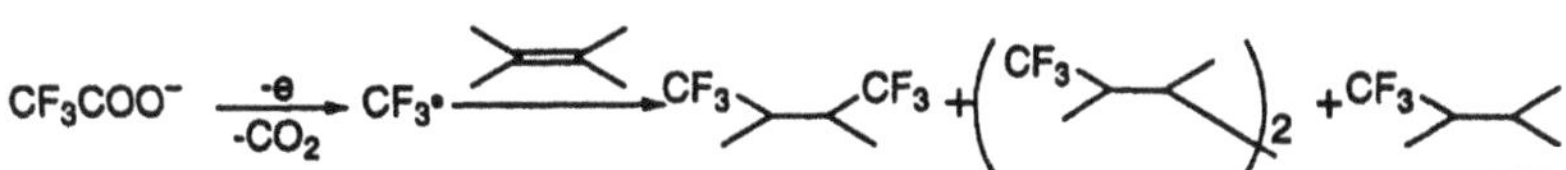

(68)

Recently, anodic generation of the CF_3 radical and its trapping reactions have been systematically studied by Uneyama and his coworkers (Scheme 24).[113–119] This work has demonstrated that trifluoromethyl radicals can be generated almost quantitatively by the oxidation of TFA at 0 °C in an aqueous MeOH/Pt system using a divided cell.[115] Also, it has been found that additions of the trifluoromethyl radical to electron-deficient olefin can be controlled by the choice of current density, reaction temperature, and olefin substituents. Notably, anodic trifluoromethylation of fumaronitrile leading to **31** (Scheme 25) is greatly affected by the reaction temperature. The

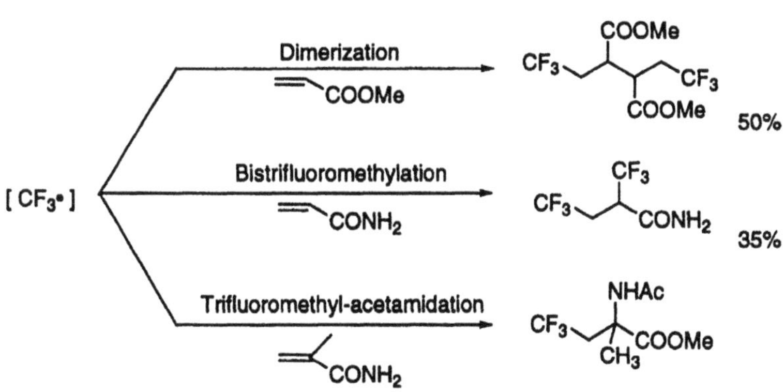

Scheme 24.

desired hydrotrifluoromethylation proceeds exclusively at ca. 55 °C while simple hydrogenation of fumaronitrile predominates at ca. 0 °C. Both anodic and cathodic processes are involved in this reaction as shown in Scheme 25.[116] Furthermore, trifluoromethylation of enolizable active-methylene compounds and enolacetates has been achieved (Eq. 69).[118]

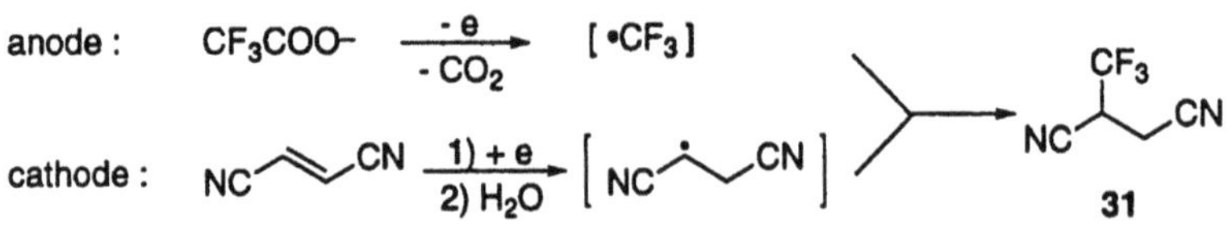

31

Scheme 25.

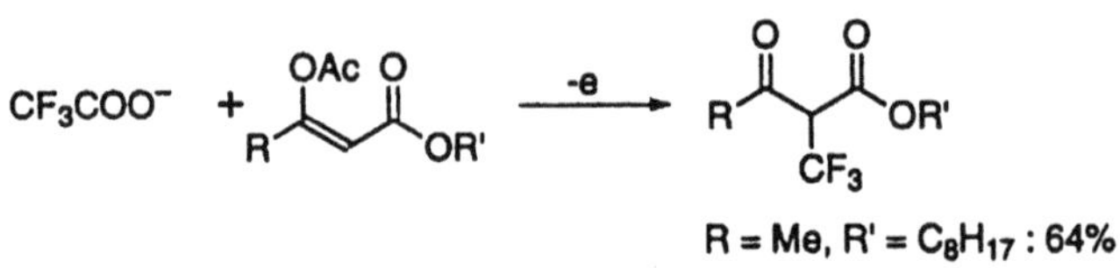

(69)

Kolbe electrolysis of trifluoromethylated carboxylic acids has been shown to be a versatile method for preparing target substances which contain the CF_3 group. Trifluoromethylation through crossed Kolbe coupling processes also has been reported as shown in Eq. 70.[120] Seebach and Renaud have prepared new types of trifluorometylated chiral building blocks from enantiomerically pure 3-hydroxy-4,4,4-trifluorobutyric acid (Scheme 26) by use of this methodology.[121]

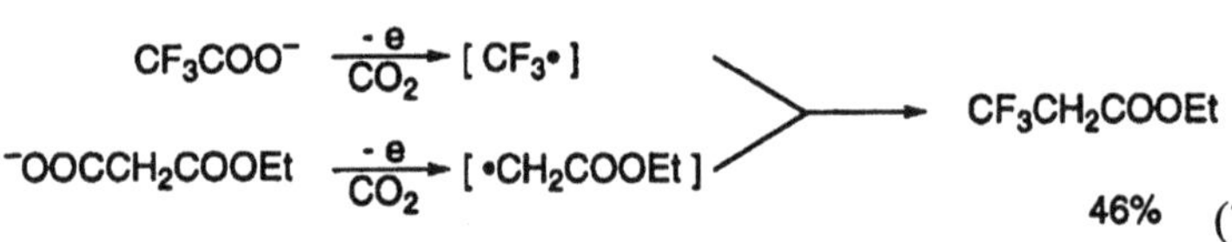

(70)

These reactions are notable because α-branched carboxylic acids usually do not undergo efficient Kolbe coupling reactions. Highly efficient homo- and crossed-coupling reactions have been achieved in a similar manner by using trifluoromethylated carboxylic acids as shown in Scheme 27 and Eq. 71.[122,123] It is worth mentioning that protection of the hydroxy groups in acids **32** is not necessary.

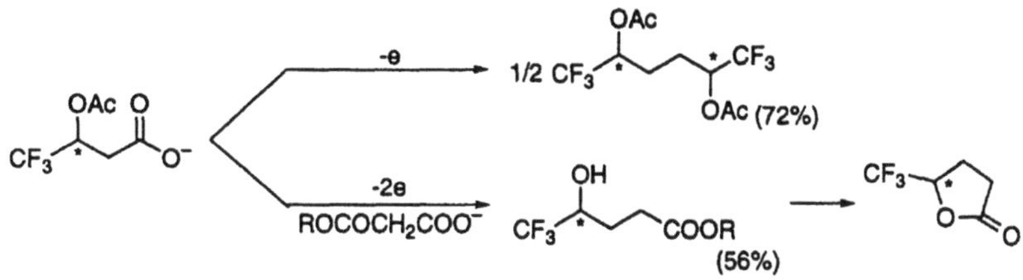

Scheme 26.

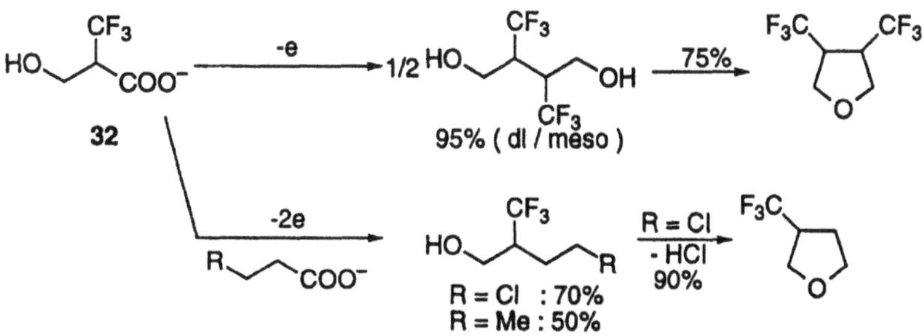

Scheme 27.

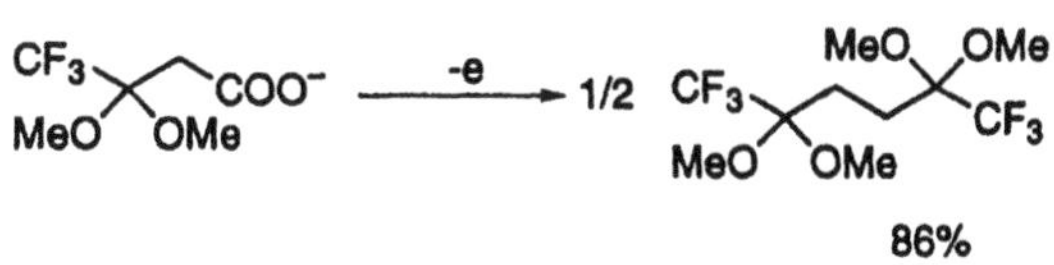

(71)

It is well-known that the anodic trifluoroacetoxylation of benzene derivatives is a useful method for the preparation of phenol derivatives (Eq. 72). Schäfer and his co-workers have successfully achieved CH-functionalization of various hydrocarbons by anodic oxidation in 0.05 M Bu_4NPF_6/CH_2Cl_2 containing 20% TFA and 4% $(CF_3CO)_2O$ as shown in Eq. 73.[124]

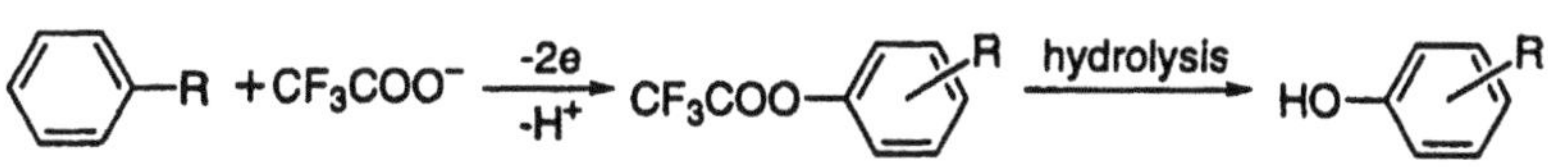

(72)

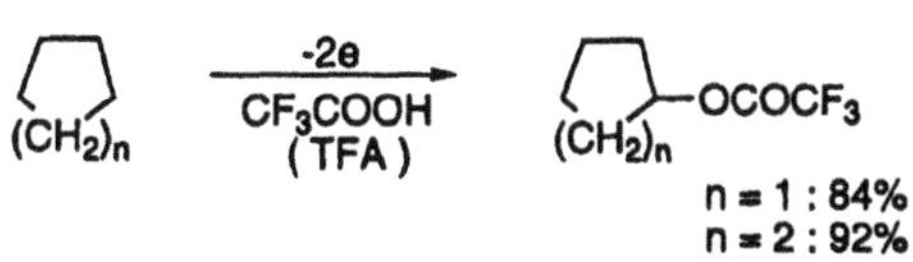

(73)

8.2. Anodic Oxidation of Trifluoromethanesulfinic Acid

In contrast to trifluoroacetate, the trifluoromethanesulfinate ion can be oxidized at a much less positive potential. Quite recently, Ignat'ev and co-workers have found that anodic oxidation of sodium trifluoromethanesulfinate results in generation of the trifluoromethyl radical, which can be trapped by arenes and followed by reoxidation and deprotonation, provides trifluoromethylated arenes in ca. 50% yield as shown in Eq. 74.[125]

$$CF_3SO_2^- \xrightarrow{-e} CF_3SO_2\bullet \xrightarrow[-SO_2]{} CF_3\bullet \xrightarrow{ArH} [CF_3ArH]\bullet \xrightarrow{-e,\,-H^+} ArCF_3 \quad (74)$$

$E_{1/2} = 1.18V$ *vs* SCE (glassy carbon anode)

$ArCF_3$ ~50%

ArH = benzene, toluene xylene, naphthalene

9. GENERAL ASPECTS OF ANODIC FLUORINATION

Another approach to the synthesis of fluorinated substances is through the use of direct fluorination. As mentioned in the Introduction, fluorinated organic compounds are not naturally occurring. Therefore, fluorine atom(s) must be introduced into other available substances at an appropriate stage in the preparative procedure. Of course, fluorine gas is useful for direct fluorination of organic substances. However, fluorination using fluorine gas is rather difficult to precisely control owing to the extremely high reactivity of fluorine gas.[126,127] This stands in sharp contrast to well-established halogenation reactions using halogens other than fluorine. In place of fluorine gas, anhydrous HF has been widely used for the synthesis of fluorinated organic substances.

Fluorinated organic substances are classified into two groups, perfluoro compounds and partially fluorinated compounds. Substances in the former class are widely utilized as functional materials while those in the latter family find biological uses as pharmaceuticals and agrochemicals.

Electrochemical fluorination was initially applied to produce perfluorinated organic substances in processes which convert all C–H to

C–F bonds. More than four decades ago, Simons and his co-workers developed procedures for electrochemical perfluorination of organic compounds using anhydrous liquid HF as solvent and a nickel anode.[128] Other electrochemical perfluorinations in a KF·2HF melt at a carbon anode have been used for the preparation of perfluorinated low-molecular-weight organic substances.[129] Both of these processes are now used commercially.

On the other hand, the study of electrochemical partial fluorination reactions (selective electrochemical fluorination) is a rather new field and methodology for this purpose is not well developed because of low reaction selectivities, the low nucleophilicity of fluoride ions, and competitive passivation of the anode.[130,131] So far, chemical methods using various reagents such as F_2, $FClO_3$, CF_3OF, XeF_2, Et_2NSF_3(DAST), and *N*-fluoropyridinium triflates have been employed in the partial fluorination processes.[132,133] However, these reagents are hazardous, difficult to handle, or very costly. In the next section, several recent examples of successful selective electrochemical fluorinations of organic compounds are described.

10. SELECTIVE ANODIC FLUORINATION

10.1. Historical Background

In 1953, anodic fluorination of 1,1-diphenylethylene was reported by the Schmidts (Eq. 75).[134] This is the first example of selective electrochemical fluorination. The procedure developed employed KHF_2/AcOH, anhydrous HF, MeCN, EtOH, AcOH containing HF,[135] or even AgF as the fluorine source.[136] However, these electrolytic systems are not always effective and thus are not widely applicable.

$$Ph_2C{=}CH_2 \xrightarrow[KHF_2/AcOH]{-2e} \underset{60\%}{Ph_2CFCH_2F} \qquad (75)$$

In contrast, Rozhkov and his co-workers found that $Et_3N{\cdot}HF$ and $Et_4NF{\cdot}3HF$ are highly effective for conducting selective fluorination reactions, particularly for nuclear fluorination of aromatic compounds (see the following section).[137]

10.2. Anodic Fluorination of Aromatic Compounds

In 1970, Rozhkov and his co-workers reported the first example of anodic partial fluorination of aromatic compounds as shown in Scheme 28.[138,139] These workers found that anodic oxidation of naphthalene in acetonitrile containing $Et_4NF \cdot 3HF$ provided 1-fluoronaphthalene (**33**) along with a small amount of 1,4-difluoronaphthalene **34**. On the other hand, the use of Et_4NF instead of $Et_4NF \cdot 3HF$ led to the efficient formation of **34** solely. This effort also demonstrated that anodic fluorination of benzene gives monofluorobenzene (**35**) as a main product as shown in Eq. 76.

35 (36%) (4.7%) (76)

Since that time, anodic partial fluorination of various aromatic compounds has been accomplished by using various fluoride salts (Eqs. 77–83).[130,140,141] 9,10-Diphenylanthracene gives 9,10-difluoro-9,10-dihydro-9,10-diphenylanthracene (**36**) as the sole product in good yield under these conditions (Eq. 77). Substituted benzenes afford fluorinated products via H-substitution (Eqs. 78–80), ipso-substitution (Eq. 81), or addition (Eqs. 82 and 83).

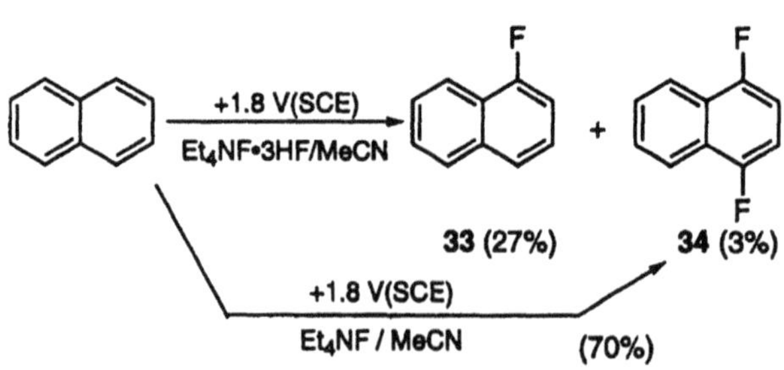

Scheme 28.

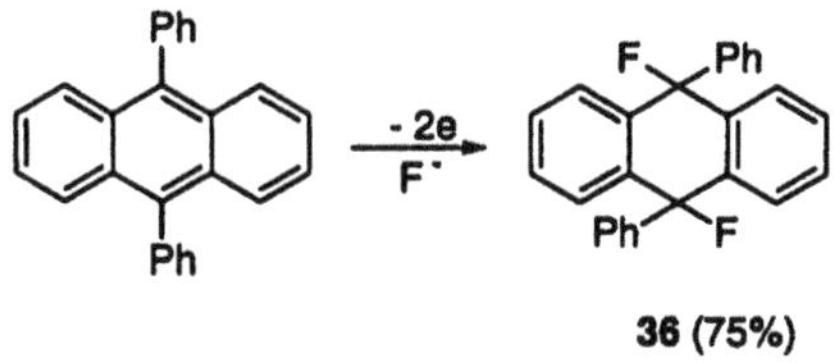

(77)

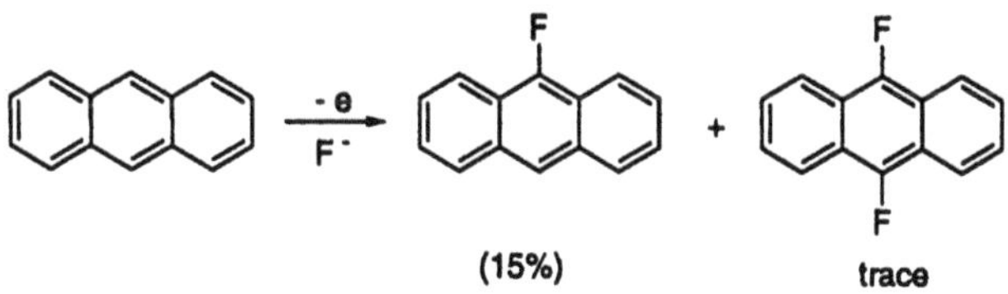

(78)

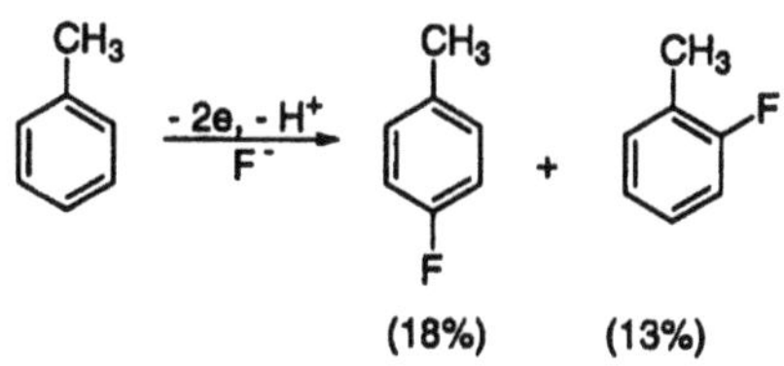

(79)

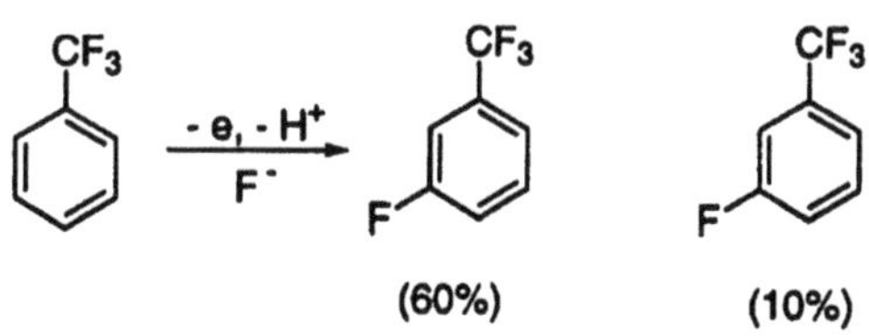

(80)

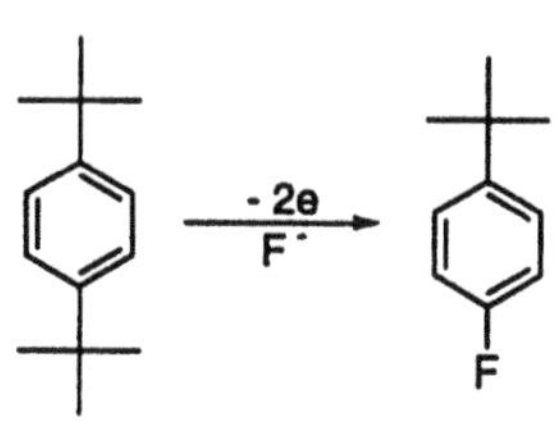

(81)

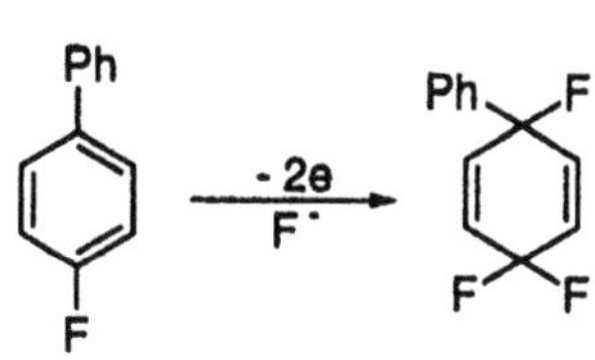

(82)

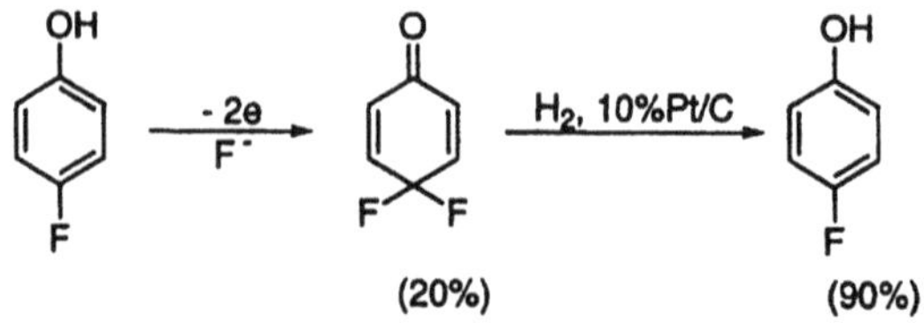

(83)

Recently, Momota and his co-workers developed a new series of fluoride salts, $R_4NF \cdot nHF$ ($R = Me$, Et, and n-Pr; $n > 3.5$), that are useful in anodic partial fluorination reactions.[142–144] These electrolytes are nonviscous liquids which have high electric conductivities and stabilities. As a result, anodic partial fluorination of arenes such as benzene, fluorobenzene, and 1,4-difluorobenzenes can be successfully carried out at high current densities by employing these fluoride salts in the absence of solvent with good to high current efficiencies (66–90%). It is known that nonconducting polymer films often form at anode surfaces when ordinary fluoride salts are used as supporting electrolytes and fluorine sources. However, such film formation is reduced considerably when these new supporting fluoride salts are used.

10.3. Anodic Benzylic Fluorination

Generally, anodic benzylic substitution reactions take place quite readily. However, anodic benzylic fluorination does not always occur as exemplified by the processes shown in Scheme 29.[145–147] The major competitive reaction is acetamidation when MeCN is used as a solvent. In contrast to these cases, triphenylmethane is selectively and efficiently monofluorinated (Eq. 84).[148]

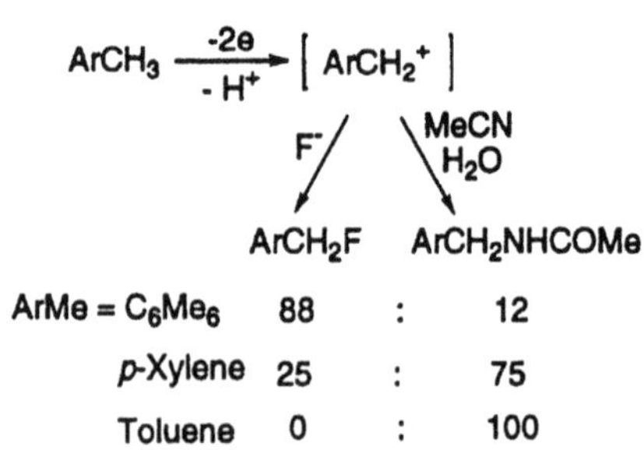

Scheme 29.

$$\mathrm{Ph_3CH} \xrightarrow[\mathrm{F^- / MeCN}]{-2e} \underset{(80\%)}{\mathrm{Ph_3CF}} \qquad (84)$$

Laurent and his co-workers found that anodic benzylic fluorination proceeds selectively when the benzylic position is substituted by electron-withdrawing groups (EWG) as shown in Eqs. 85 and 86.[149] In these cases, *p*-methoxy or *p*-chloro substituents on the benzene ring are necessary for the operation of efficient fluorination (Eq. 85). In their absence, benzylic acetamidation becomes a major reaction (Eq. 85). In later studies, this group found that the use of sulfolane instead of MeCN as solvent leads to higher yields of fluorinated products because of the absence of competing acetamidation.[150] They also were able to achieve geminal difluorination reactions by using large amounts of electricity (6 F/mol) (Eq. 87).[151]

$$\mathrm{ArCH_2COMe} \xrightarrow[\mathrm{F^- / MeCN}]{-2e,\ -H^+} \mathrm{ArCHFCOMe} + \mathrm{ArCH(NHCOMe)COMe} \qquad (85)$$

Ar = Ph:	7%	34%
p-MeOC$_6$H$_4$	69%	<1%

$$p\text{-}\mathrm{XC_6H_4CH_2{-}EWG} \xrightarrow[\mathrm{F^- / MeCN}]{-2e,\ -H^+} p\text{-}\mathrm{XC_6H_4CHF{-}EWG} \qquad (86)$$

36 ~ 73%

X = MeO, Cl
EWG = COAr, COOEt, CN, SO$_3$Et

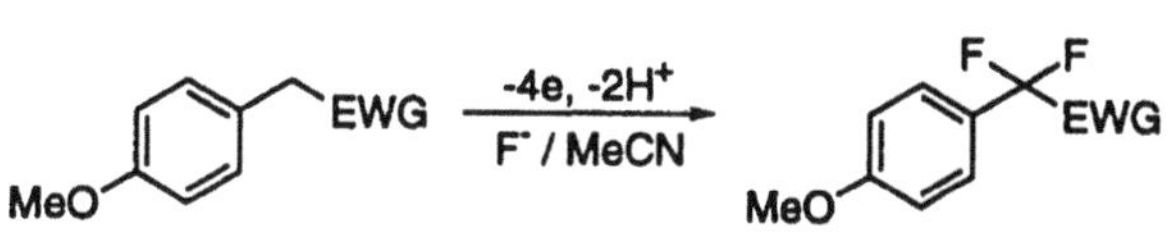

(87)

48 ~95%

EWG = COMe, COAr, COOEt
CN, SO$_3$Et

10.4. Anodic Fluorination of Olefins

Anodic oxidation of olefins in the presence of fluoride ions provides mono- and/or difluorinated products as shown in Eqs. 75, 88 and 89.[134,152] Butadiene under these conditions gives a 1:2 mixture of 1,2- and 1,4-adducts as shown in Eq. 90.[153] Anodic fluorination of vinyl sulfides such as 2-(phenylthio)styrene provides vicinal difluorides as shown in Eqs. 91 and 92.[154] At more positive potentials, a trifluorinated sulfide is directly obtained in a one-pot reaction (Eq. 91).[154]

$$\mathrm{PhCH{=}CHPh} \xrightarrow[\mathrm{F^- / MeCN}]{-2e} \mathrm{PhCHFCHPh(NHCOMe)} \quad (35 \sim 60\%) \tag{88}$$

$$\mathrm{CH_3CH{=}CH_2} \xrightarrow[\mathrm{Bu_4BF_4^- /2\%AcOH\text{-}CH_2Cl_2}]{-2e} \mathrm{CH_3CHFCH_2OAc} \quad 30\% \tag{89}$$

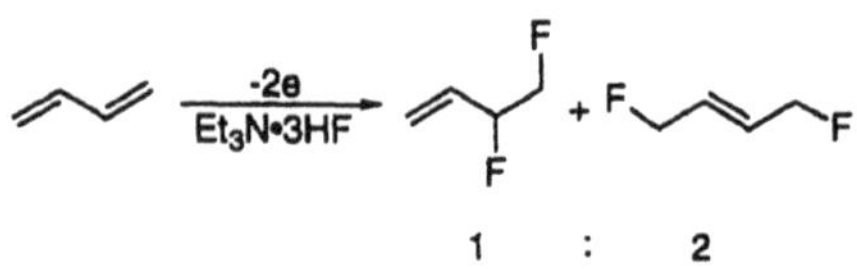

(90)

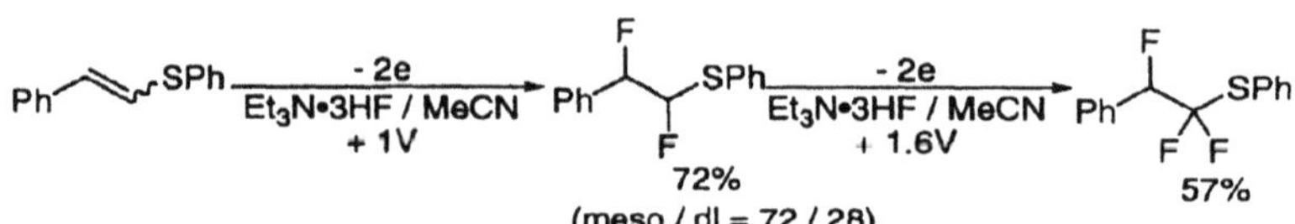

(91)

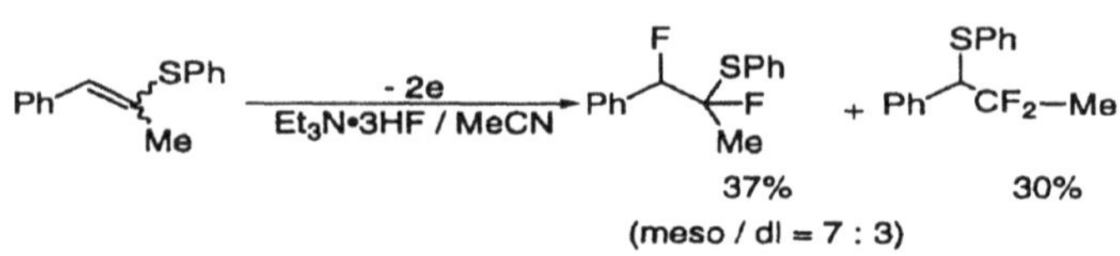

(92)

1-Phenylhexene (**27**) undergoes stereoselective difluorination and fluoroacetamidation on anodic oxidation in MeCN while the difluorination predominates in the less nucleophilic solvent CH_2Cl_2 as shown in Scheme 30.[147]

Electrochemical fluoroacetamidations of 1-alkylindene were investigated by Laurent's group. The stereoselectivities with respect to F/alkyl and F/MeCONH introduction were found to be mainly *trans* and *cis*, respectively.[155] In contrast, Yoneda and his co-workers found that anodic oxidation of cyclic unsaturated esters in $Et_3N{\cdot}5HF$ resulted

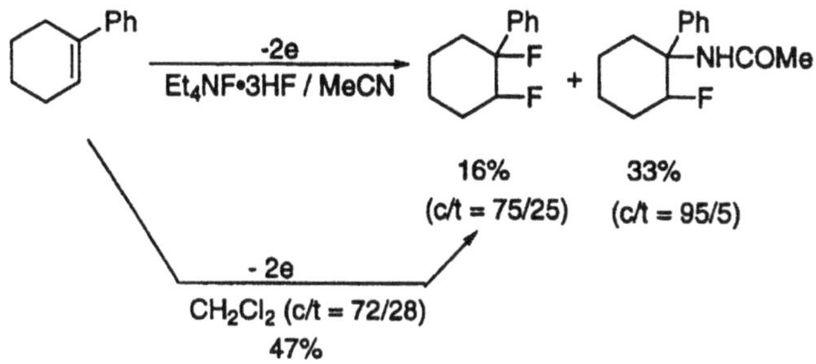

Scheme 30.

in fluorinative ring expansion to provide 2,2-difluorocycloalkane carboxylates as shown in Eq. 93.[156] Also, α-fluoroketones were prepared from α-acetoxystyrene and 1-acetoxy-3,4-dihydronaphthalene (Eqs. 94 and 95).[157,158]

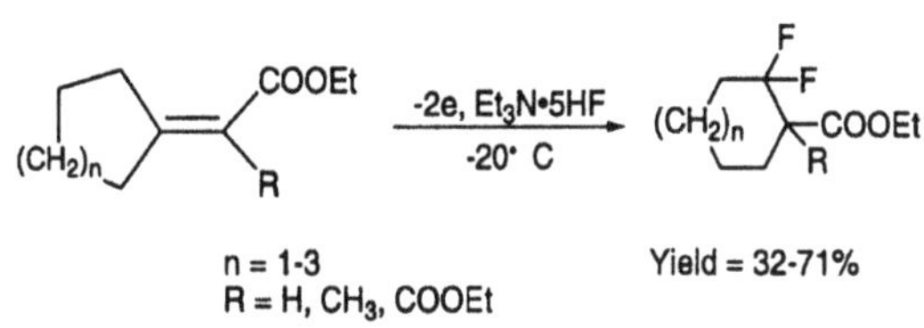

(93)

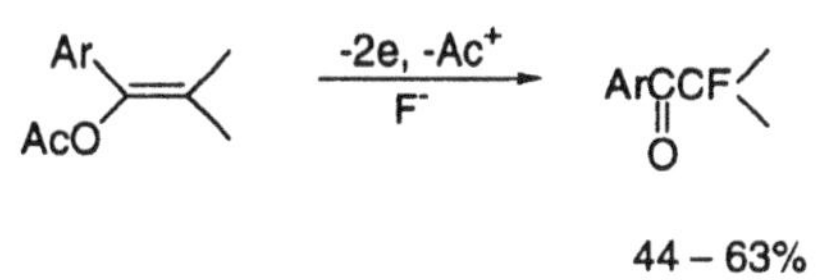

(94)

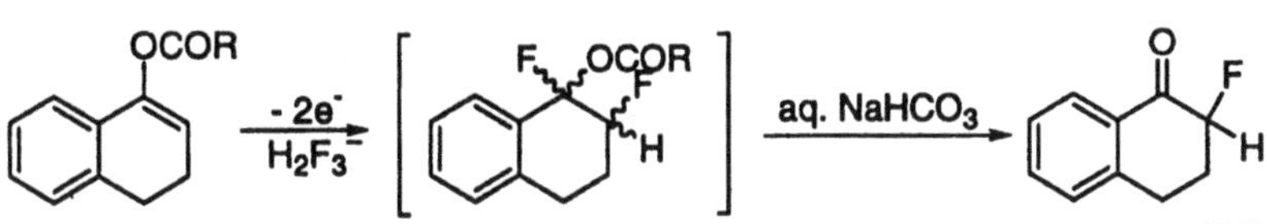

(95)

10.5. Anodic Fluorination of Carbonyl Compounds

Selective anodic fluorination reactions of aldehydes and cyclic ketones can be successfully carried in $Et_3N{\cdot}5HF$ to give the corresponding acyl fluorides and fluoroacyl fluorides in good yields (Table 12, Scheme 31).[159] In these reactions, fluoride salts such as $Et_3N{\cdot}3HF$

Table 12. Anodic Fluorination of Aldehydes

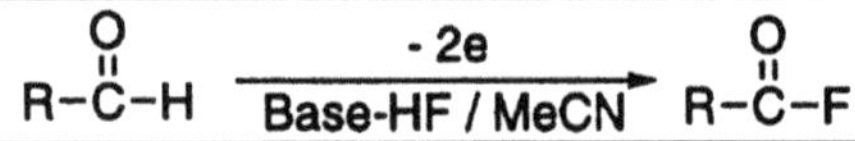

RCHO *R*	*HF-Base*	*Yield (%)*
Heptyl	Pyridine·6HF	58
Heptyl	$Et_3N\cdot3HF$	37
Heptyl	$Et_3N\cdot5HF$	85
Cyclohexyl	$Et_3N\cdot5HF$	84
t-Butyl	$Et_3N\cdot5HF$	66

lead to lower yields, reflecting the fact that this salt is discharged prior to oxidation of the carbonyl compounds.

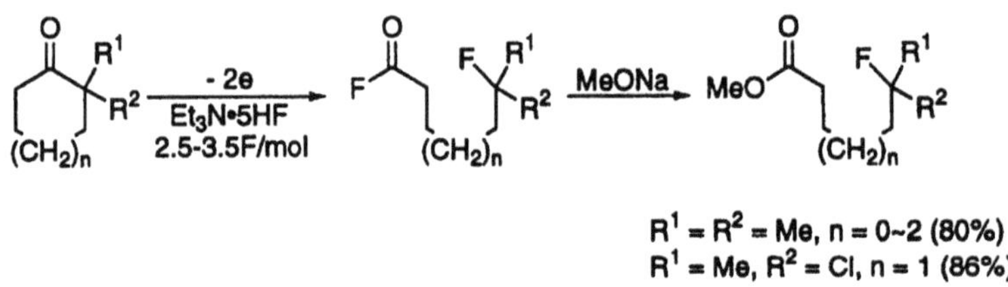

Scheme 31.

10.6. Anodic Fluorination of Chalcogeno Compounds

As described above, studies of anodic partial fluorination have focused on aromatic, benzylic and olefinic compounds. However, there have as yet been no reports on anodic partial fluorination reactions of chalcogeno compounds.

Anodic Fluorination of Organosulfur Compounds

As mentioned above, Fuchigami and his co-workers found that a trifluoromethyl group markedly promotes anodic methoxylation and acetoxylation of 2,2,2-trifluoroethyl sulfides (Scheme 14 and Eq. 96). Based on this finding, this group has explored the anodic substitution reactions of trifluoroethyl sulfides with fluoride ions.[81,160] Anodic monofluorination was first probed in detail using phenyl 2,2,2-trifluoroethyl sulfide (**37a**) as a model compound. Anodic oxidation of **37a** was carried out at constant potential in acetonitrile containing various fluorides as supporting electrolytes and fluoride ion sources by using an undivided cell. As seen in Table 13, anodic monofluorination procedes smoothly, providing tetrafluoroethyl sulfide **38a** in good yield only when $Et_3N{\cdot}3HF$ is used (run 3); the other two fluorides were not effective in this process (runs 1 and 2).

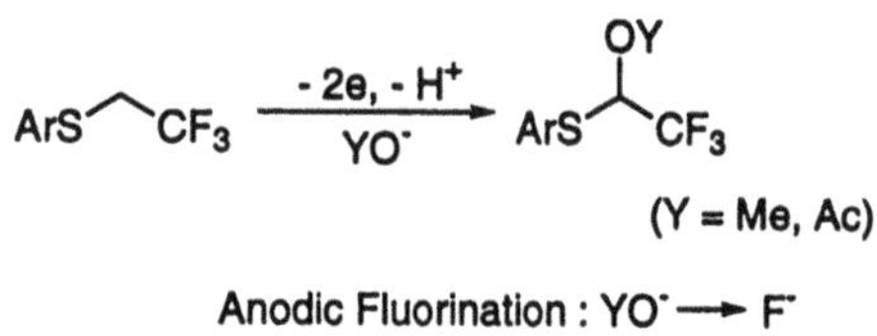

(96)

This fluorination methodology was extended to various *para*-substituted phenyl 2,2,2-trifluoroethyl sulfides, **37b–d**. These substances undergo efficient fluorination regardless of the benzene ring substituent. Fluorine is exclusively introduced at the position α to the trifluoromethyl group. Neither aromatic fluorination nor benzylic fluorination was observed (run 5, Table 13). The results are noteworthy

Table 13. Anodic Monofluorination of Aryl 2,2,2-Trifluoroethyl Sulfides **37**

$$\underset{\textbf{37}}{p\text{-}XC_6H_4SCH_2CF_3} \xrightarrow[F^-/\,MeCN]{-2e,\,-H^+} \underset{\textbf{38}}{p\text{-}XC_6H_4SCHFCF_3}$$

Run	*Sulfide X*	*Supporting Electrolyte*	*Anodic Potential (V vs. SSCE)*	*Charge Passed (F/mol)*	*Product Yield (%)*	
1	H (**37a**)	$Py{\cdot}(HF)_n$	+2.0	4.1	(**38a**)	0
2	H (**37a**)	$Bu_4NF{\cdot}3H_2O$	+2.0	1.9	(**38a**)	0
3	H (**37a**)	$Et_3N{\cdot}3HF$	+1.9	3.2	(**38a**)	62
4	Cl (**37b**)	$Et_3N{\cdot}3HF$	+2.0	7.2	(**38b**)	65
5	Me (**37c**)	$Et_3N{\cdot}3HF$	+2.1	8.2	(**38c**)	51
6	MeO (**37d**)	$Et_3N{\cdot}3HF$	+1.7	6.0	(**38d**)	56

Table 14. Anodic Monofluorination of Fluoroalkyl Phenyl Sulfides **37**

$$\underset{\mathbf{37}}{PhSCH_2Rf} \xrightarrow[Et_3N\cdot 3HF]{-2e,\ -H^+} \underset{\mathbf{38}}{PhSCHFRf}$$

Run	Sulfide Rf	Anodic Potential (V vs. SSCE)	Charge Passed (F/mol)	Product Yield (%)
1	CF_3 (**37a**)	+1.9	3.2	(**38a**) 62
2	C_3F_7 (**37e**)	+2.3	3.0	(**38e**) 67
3	CHF_2 (**37f**)	+2.3	4.0	(**38f**) 53
4	CH_2F (**37g**)	+2.1	2.7	(**38g**) 60
5	CF_2Cl (**37h**)	+2.0	5.0	(**38h**) 46
6	CH_2Cl (**37i**)	+1.8	3.0	(**38i**) 30

because nucleophilic substitution at the position α to the trifluoromethyl group is usually quite difficult to achieve. Moreover, products **38** are not readily preparable by other methods.

Studies of the anodic fluorination process were extended to various fluoroalkyl phenyl sulfides (**37e–i**).[82,160] The results of this effort are summarized in Table 14. As shown, anodic fluorination proceeds to afford the corresponding α-monofluorinated products **38** regardless of the nature of the fluoroalkyl groups. A strong electron-withdrawing

Table 15. Anodic Monofluorination of Alkyl Phenyl Sulfides **39**

$$\underset{\mathbf{39}}{PhSCH_2R} \xrightarrow[Et_3N\cdot 3HF]{-2e,\ -H^+} \underset{\mathbf{40}}{PhSCHFR}$$

Run	Sulfide R	Solvent	Product Yield (%)
1	H (**39a**)	MeCN	23
2	H (**39a**)	THF	52
3	H (**39a**)	DME	57
4	H (**39a**)	Dioxane	47
5	Me (**39b**)	MeCN	18
6	Me (**39b**)	THF	45

perfluoroalkyl group, such as a heptafluoropropyl, promotes anodic fluorination (run 2). Interestingly, weaker electron-withdrawing groups such as difluoromethyl and monofluoromethyl similarly promote fluorination (runs 3 and 4). Therefore, it appears that the electron-withdrawing ability of the fluoroalkyl group does not affect the efficiency of anodic monofluorination reactions of sulfides. These results are quite different from those observed in studies of anodic methoxylation (see above).

It is notable that the enhancing effect of a fluorine atom on the anodic fluorination process is much more pronounced than that of a chlorine atom, although their electronegativities are similar (Table 14, runs 1, 5 and 4, 6). Therefore, the effect of a fluorine atom is quite specific.

Further, it was found that even simple alkyl phenyl sulfides **39** devoid of an electron-withdrawing group undergo fluorination to provide monofluorination products **40**, although the yields are not high (Table 15, runs 1 and 5).[160] Since the nucleophilicity of fluoride ions is known to be greatly affected by solvents, the solvent for these electrolytic reactions was changed to achieve higher yields. Thus, the yields were approximately doubled when ethereal solvents such as tetrahydrofuran (THF), dimethoxyethane (DME), and dioxane instead of acetonitrile were used in this process (runs 2–4, 6). Products **40** are known to be useful starting materials for the preparation of monofluorovinyl sulfides.

The electrochemical reaction–chemical reaction–electrochemical reaction–chemical reaction (ECEC) mechanism is well established for anodic methoxylation reactions of amines, amides, and carbamates. It was confirmed that the anodic methoxylation of fluoroalkyl sulfides also proceeds with by way of a typical ECEC mechanism as shown in Scheme 32.[69]

Anodic fluorinations of fluoroalkyl sulfides **37** also appear to proceed by a similar mechanism. However, the effects of both substituents (X and Rf groups in Scheme 32) on this process are totally different from those observed for anodic methoxylation. Therefore, the mechanism for anodic fluorination could very well differ from that for anodic methoxylation.

Fuchigami, Konno, and their co-workers observed interesting phenomena in their study of this process. Namely, anodic fluorination of

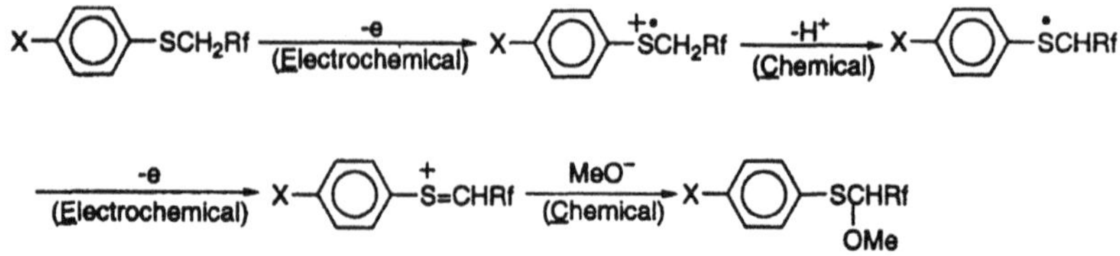

Scheme 32.

37g in methanol provides α-methoxylated product **13** instead of α-fluorinated product **38g** in a good yield (Scheme 33). As reported previously, **13** is not obtained under conventional anodic methoxylation conditions, and this result is attributed to very slow deprotonation of the cation radical intermediate due to the weak electron-withdrawing ability of the CH_2F group. Thus, this marked promotion effect of $Et_3N{\cdot}3HF$ on the α-methoxylation of **37g** cannot be explained by the conventional ECEC mechanism.

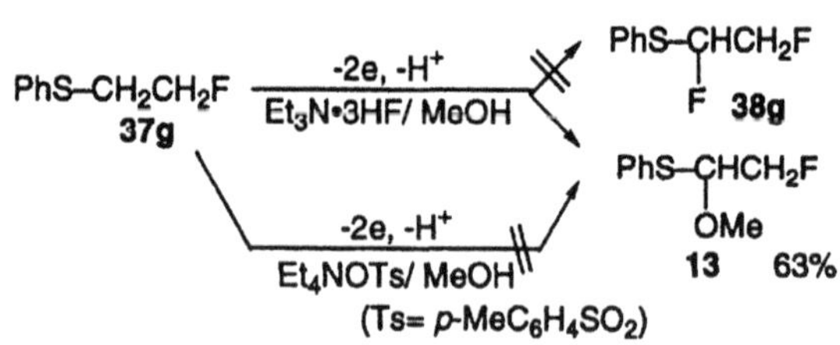

Scheme 33.

From these results, it appears that this anodic fluorination proceeds by way of a Pummerer-type mechanism via the fluorosulfonium cation **K**, as shown in Scheme 34. In this pathway, the cation radical **J** of the sulfide is trapped by fluoride ion. This step suppresses side reactions of **J** (such as dimerization and nucleophilic attack on an aromatic ring) even when deprotonation of this intermediate is slow. Since fluoride ion is a much weaker nucleophile than methoxide, it is reasonable that methoxylation predominates in methanol. Thus, efficient α-fluorination in acetonitrile and α-methoxylation in methanol of **37g** in the presence of $Et_3N{\cdot}3HF$ can both be explained by assuming the common intermediate **K**.

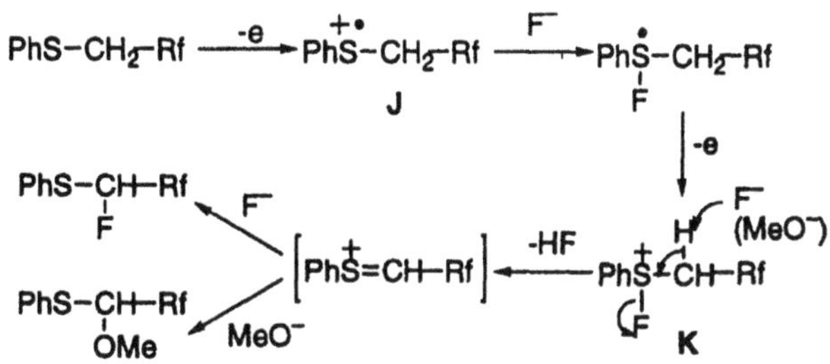

Scheme 34.

Anodic fluorination is a widely applicable methodology.[161] Thus, α-monofluorination of sulfides bearing electron-withdrawing substituents other than fluoroalkyl is also successful, as shown by the examples included in Table 16. Even though α-thio-substituted esters **41b, 41c,** and **41d** each have multiple positions susceptible to substitution by fluorine, highly regioselective fluorination takes place with fluorine being introduced exclusively α to the ester group. No fluorination of the *para*-tolyl, benzyl, or heptyl groups is observed. Fluorination of malonic ester **41i** provides the expected fluorinated product **42i** (run 9). However, diketone **41h** gives fluorinated monoketone **42g** instead of the expected diketone **42h** under these conditions (run 8). Fluorination is also successful via α-(phenylthio)-substituted cyclic ketones **43a** and **43b** and a lactone analogue (Eqs. 97–99).[161]

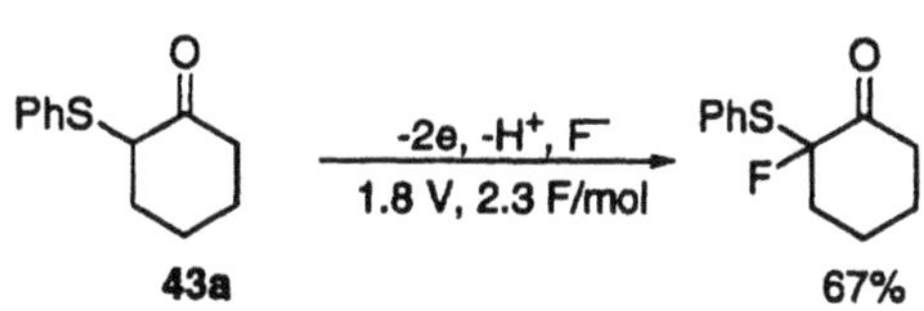

(97)

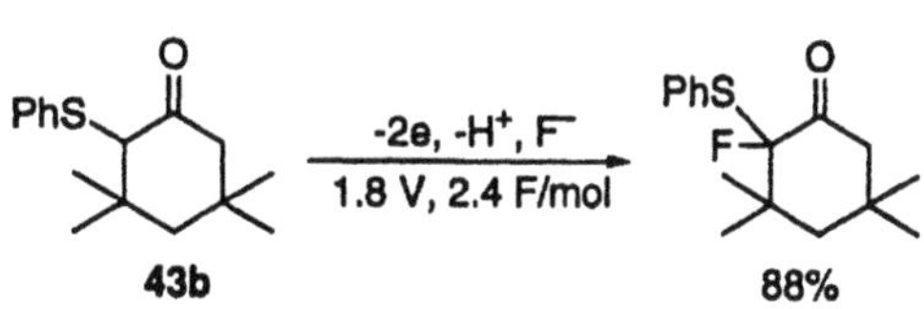

(98)

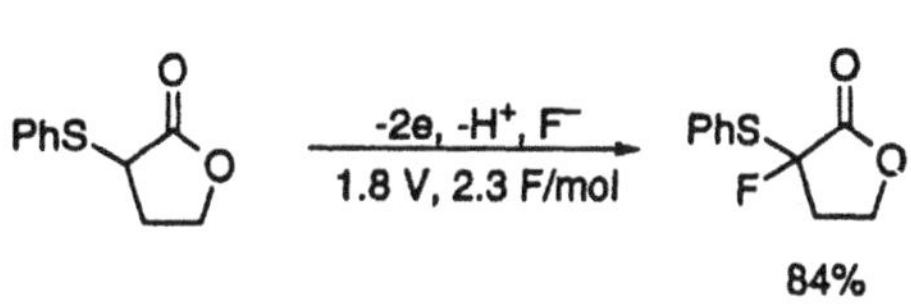

(99)

Table 16. Anodic Monofluorination of Sulfides Bearing Electron-Withdrawing Groups

$$\underset{\mathbf{41}}{R^1SCHR^2R^3} \xrightarrow[Et_3N\cdot 3HF]{-2e,\ -H^+} \underset{\mathbf{42}}{R^1SCFR^2R^3}$$

Run	*Sulfide* R^1	R^2	R^3		*Anodic Potential (V vs. SSCE)*	*Charge Passed (F/mol)*	*Product Yield (%)*	
1	Ph	H	COOEt	(**41a**)	1.6	2.5	(**42a**)	76
2	*p*-MeC_6H_4	H	COOEt	(**41b**)	1.6	2.1	(**42b**)	78
3	$PhCH_2$	H	COOEt	(**41c**)	2.1	5.0	(**42c**)	44
4	C_7H_{15}	H	COOEt	(**41d**)	2.1–2.3	16.1	(**42d**)	70
5	Ph	H	CN	(**41e**)	1.7	5.0	(**42e**)	75
6	Ph	H	COPh	(**41f**)	1.5	5.0	(**42f**)	55
7	Ph	H	COMe	(**41g**)	1.6–1.8	7.6	(**42g**)	80
8	Ph	COMe	COMe	(**41h**)	1.7	3.0	(**42g**)	55
9	Ph	COOEt	COOEt	(**41i**)	2.0	15.4	(**42i**)	77
10	Ph	Me	COOEt	(**41j**)	1.8	2.9	(**42j**)	83
11	Ph	Cl	COOEt	(**41k**)	2.2	2.3	(**42k**)	66
12	Ph	H	CONHPr	(**41l**)	1.7	2.8	(**42l**)	88
13	Ph	H	$PO(OEt)_2$	(**41m**)	1.9	2.7	(**42m**)	84

Brigaud and Laurent independently uncovered a similar anodic fluorination process.[162] They used a divided cell for fluorination of α-benzoyl sulfide **41f**. On the contrary, Fuchigami and his co-workers found that sulfides **41f–h** and **43a–b**, all having reducible carbonyl groups, undergo selective anodic fluorination even in an undivided cell. Because of the presence of acidic protons, $Et_3N{\cdot}3HF$ is discharged at a relatively early stage of the electrolysis, and thus the carbonyl groups are not reduced. However, a divided cell is necessary to perform efficient fluorination in the case of **41k** because of the presence of an easily reducible C–Cl bond (Table 17 and run 11 of Table 16).

Simonet and his co-workers similarly performed regioselective anodic α-monofluorination of alkyl aryl sulfides, having an electron withdrawing group on the aromatic ring, in an $Et_3N{\cdot}3HF$/MeCN solution.[163]

The diastereoselective anodic fluorination of α-phenylsulfenyl esters via intramolecular asymmetric induction has been studied using various chiral auxiliaries.[164] Of these chiral auxiliaries, the 8-phenylmethyl group gives the best diastereoselectivities. Diastereoselectivity is also affected by the supporting electrolyte and it is found that $Et_4NF{\cdot}3HF$ leads to better selectivity relative to $Et_3N{\cdot}3HF$ or $Et_3N{\cdot}2HF$ as shown in Scheme 35.

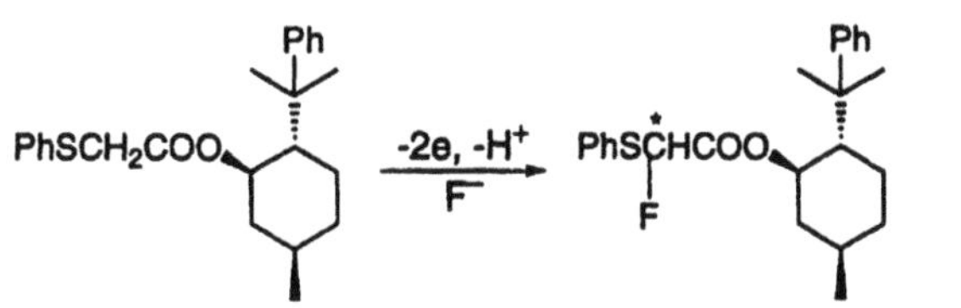

$Et_3N{\cdot}3HF$	: 69% yield, 16% d.e.
$Et_4NF{\cdot}2HF$	: 55% yield, 18% d.e.
$Et_4NF{\cdot}3HF$	: 37% yield, 28% d.e.

Scheme 35.

Anodic Fluorination of Organoselenium Compounds

Anodic monofluorination of selenides (**44a** and **44b**) bearing electron-withdrawing cyano and ester groups can be performed in a divided cell (Table 17).[165] However, reaction of α-selenoamide **44c** provides no α-fluorinated product. When a divided cell equipped with a sintered glass diaphragm is used, the desired fluorinated product **45c** is obtained in low yield (run 4). In this case, the starting diphenyl diselenide **44c** and **45c** are detected in the catholyte. These results suggest that anodic oxidation of **44c** leads to generation of a relatively stable cationic intermediate which permeates from the anodic to the cathodic chamber where it is subjected to cathodic reduction to regenerate **44c**. In order to avoid this permeation process, an anion-exchange membrane is used as a diaphragm and this variation results in a satisfactory yield of **45c** (run 5).

It should be noted that selenides **44d** and **44e** bearing an additional electron-withdrawing group (e.g., ester or chlorine atom) provide fluorinated selenides **45d** and **45e** in moderate yields, while α-alkylated α-seleno-substituted ester **44f** gives only a trace amount of the fluorinated product **45f** (run 8).

According to these results, the anodic fluorination reactions of selenides seem to involve fluoroselenonium ions **L** as a key intermediate in a manner similar to that seen in the case of sulfides. In fact, a

Table 17. Anodic Monofluorination of Selenides

$$\underset{\mathbf{44}}{PhSeCHR^1R^2} \xrightarrow[Et_3N\cdot 3HF\ /\ MeCN]{-2e,\ -H^+} \underset{\mathbf{45}}{PhSeCFR^1R^2}$$

Run	Selenide R^1	Selenide R^2		Type of Cell[a]	Anodic Potential (V vs. SSCE)	Charge Passed (F/mol)	Product	Yield (%)
1	H	CN	(**44a**)	U	1.6–1.8	6.6	(**45a**)	71
2	H	COOEt	(**44b**)	U	1.5	5.8	(**45b**)	70
3	H	$CONH_2$	(**44c**)	U	1.6	4.0	(**45c**)	0
4	H	$CONH_2$	(**44c**)	Dg	1.6–1.7	2.4	(**45c**)	19
5	H	$CONH_2$	(**44c**)	Di	1.5–1.6	3.5	(**45c**)	60
6	COOEt	COOEt	(**44d**)	Di	1.8	2.7	(**45d**)	55
7	Cl	COOEt	(**44e**)	Di	2.3	3.0	(**45e**)	65
8	*n*-Hep	COOEt	(**44f**)	Di	1.8	3.6	(**45f**)	trace

Note: [a]U, undivided; Dg, divided with a sintered glass; Di, divided with anion-exchange membrane.

fluoroselenonium ion is detected by mass spectrometry in the anolyte of anodic oxidation of simple methyl phenyl selenide. Isolation of **L** as a stable salt was attempted. After the electrolysis of α-alkylated α-seleno-substituted esters **44f** and **44g**, sodium tetraphenylborate was added to the anolyte. Interestingly, diaryl selenide **46f** was obtained from **44f** and the expected salt **47f** was not isolated. It is notable that *para*-chlorophenyl selenide **44g** reacts to generate *para*-chlorophenyl phenyl selenide **46g** solely and **46f** is not obtained (Scheme 36). Therefore, in this reaction, the aryl ring attacks the selenonium cation to form the hypervalent intermediate **L** from which characteristic ligand coupling occurs providing selenides **46f** and **46g**, as illustrated in Scheme 36.[166]

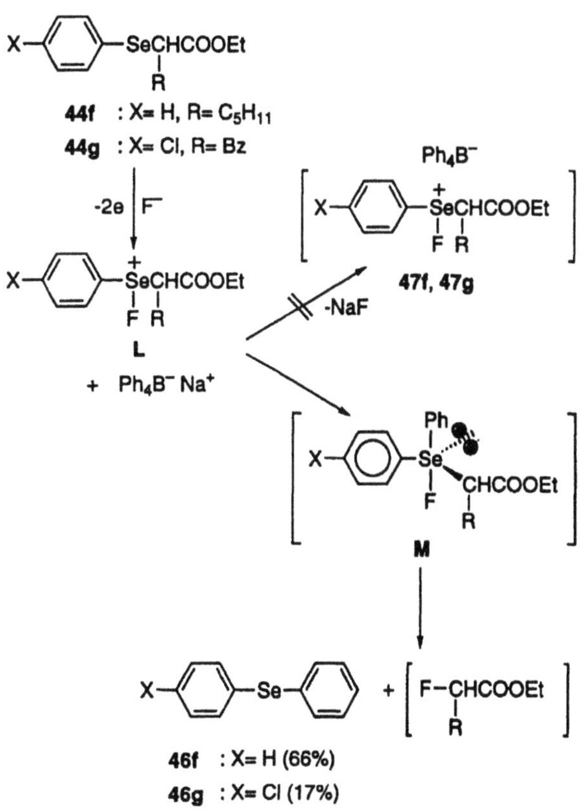

Scheme 36.

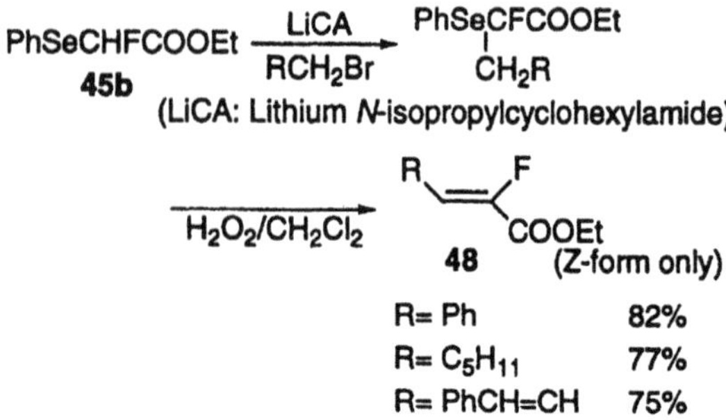

Scheme 37.

By using the α-fluoro α-seleno-substituted ester **45b** prepared in this fashion, a highly stereoselective synthesis of α-fluoro α,β-unsaturated ester **48** was achieved, as shown in Scheme 37. Substances of this type are useful for the preparation of monofluorinated retinoids, insect sex pheromones, and pyrethroids.[105]

Anodic Fluorination of Organotellurium Compounds

Anodic oxidation of organotellurium compounds in the presence of fluoride ion results in difluorination at the tellurium atom selectively, in excellent yields and with high current efficiencies (Eq. 100).[167] This is the first example of anodic fluorination of tellurides. Even a trifluoroethyl telluride does not give an α-fluorotelluride but rather reacts by way of Te–F bond formation.

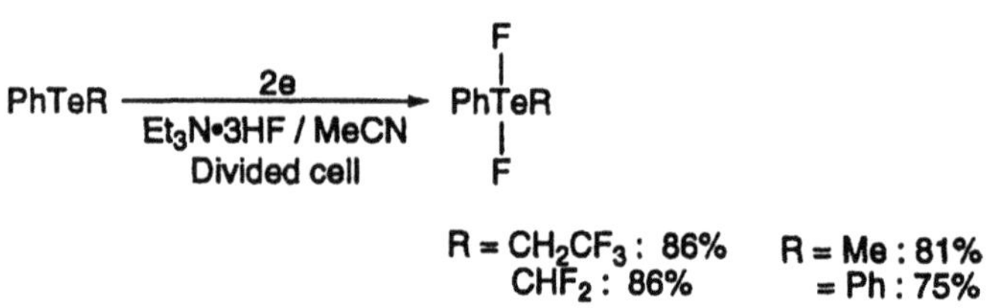

(100)

10.7. Anodic Fluorination of Other Heteroatom Compounds

Anodic oxidation of tetraalkylsilanes in the presence of fluoride ions provides the corresponding fluorosilanes derived from cleavage of the C–Si bond.[168] The proposed intermediates in these processes are the pentacoordinate silyl radicals, which eliminate the most stable alkyl radical to give the corresponding fluorosilane as shown in Eq. 101.

$$R_3SiR' \xrightarrow[F^-]{-e} R_3\dot{Si}(F)(R') \xrightarrow[-\bullet R']{} R_3SiF \qquad (101)$$

Anodic oxidation of organic compounds containing group 15 elements in the presence of fluoride ions provides the corresponding fluorinated products (Eqs. 102 and 103).[169–174] Fluorination occurs at the heteroatoms selectively.

$$R_3Z \xrightarrow[F^-]{-2e} R_3ZF_2 \qquad (Z = P, Sb, As) \qquad (102)$$

$$R_2PH \xrightarrow[-F^-]{-2e,\,-H^+} R_2PF \xrightarrow[2F^-]{-2e} R_2PF_3 \xrightarrow{H_2O} R_2P(O)F \;\; 83\text{-}90\% \qquad (103)$$

10.8. Anodic Fluorination of Heterocyclic Compounds

Many heterocyclic compounds have unique biological activities. Also, introduction of fluorine atom(s) into organic substances sometimes markedly enhances or dramatically changes their biological activities. Therefore, partially fluorinated heterocycles are the focus of much biological interest. However, limited examples of selective anodic fluorination reactions of heterocycles have been reported as shown in Eqs. 104–108.[175–179] These processes are limited to only nitrogen- and oxygen-containing heterocycles, and the yields are generally quite low. In addition, no successful anodic fluorination reaction of sulfur-containing heterocycles has been reported to date.

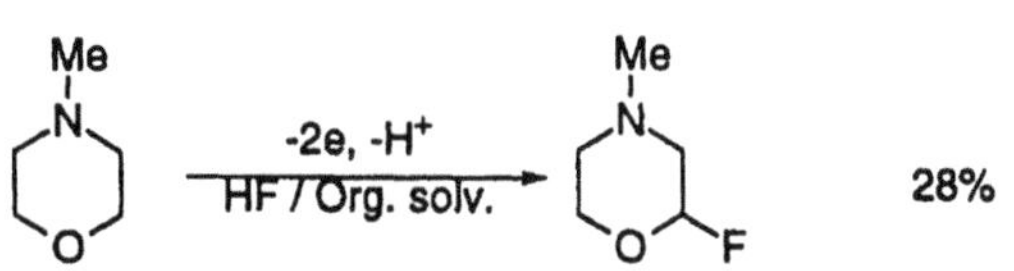

(104)

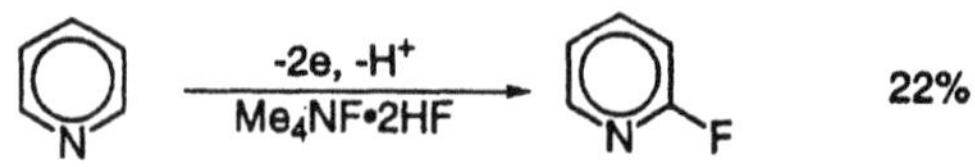

(105)

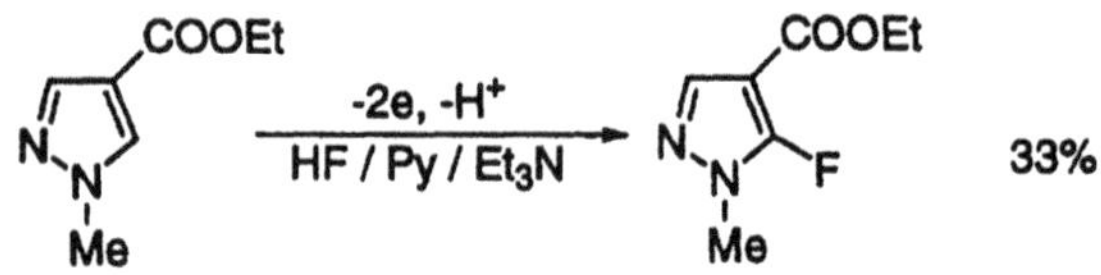

(106)

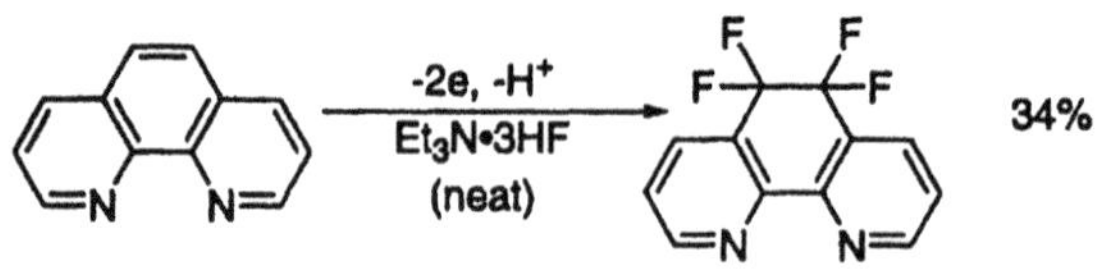

(107)

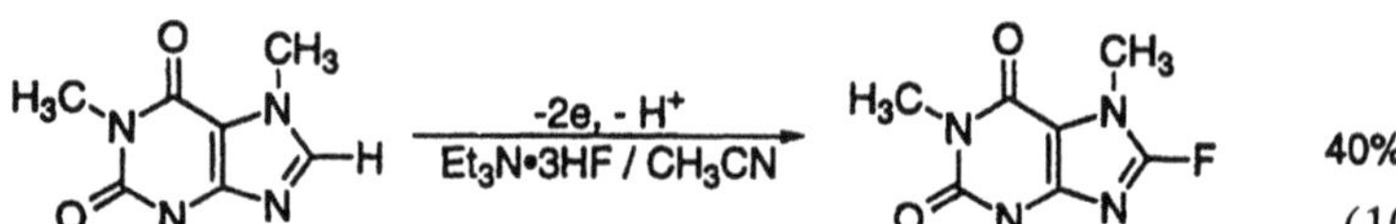

(108)

Fuchigami and his co-workers have developed conditions for highly selective anodic fluorination reactions of various heterocyclic compounds.[180]

Anodic Monofluorination of Heterocycles

As shown in Eq. 109, 4-thianones provide the corresponding monofluorination products in reasonable yields and with moderate to high diastereoselectivity.[181]

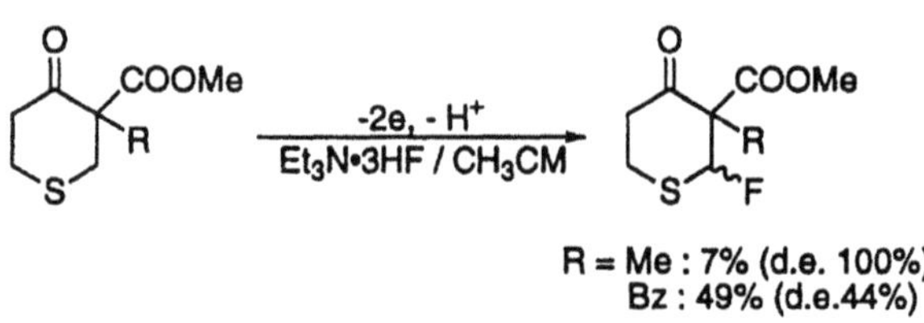

Anodic monofluorination of 4,4-dimethyl-2-ethoxycarbonyl-3-thiolanones **49** proceeds smoothly without passivation of the anode in $Et_3N{\cdot}3HF/MeCN$ to provide the corresponding 6-fluorinated products **50** in good yields (Scheme 38).[182] When R in **49** is hydrogen, fluorination takes place at the 2-position selectively.

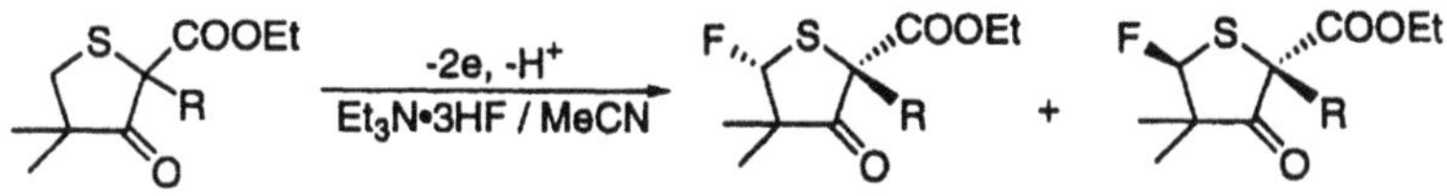

49 **50**

50a: R= Me (74%) (trans/cis: 74/25)
50b: R= Bzl (71%) (trans/cis: 61/39)
49 (R= H) ⟶ **49** (R= F: 60%)

Scheme 38.

Highly regioselective anodic monofluorination of 2-substituted 1,3-dithiolan-4-ones **51** and 1,3-oxathiolan-4-ones **52** can be successfully carried with the novel supporting electrolyte $Et_4NF{\cdot}4HF$ developed by Momota while a conventional supporting electrolyte, $Et_3N{\cdot}3HF$ gives poor yields and/or low current efficiencies as a result of severe passivation of the anode during the electrolysis (Scheme 39).[183,184]

In the case of 2-aryl substituted 1,3-dithiolan-4-ones and 1,3-oxathiolanones, benzylic fluorination is not observed regardless of the nature of the supporting electrolytes although in general anodic benzylic substitution easily takes place. Therefore, the regioselectivity of this fluorination is not controlled by the stability of the cationic intermediates **O** and **O′** as the fluorinated products are formed via the less stable intermediates **O** (Scheme 40). Instead, the regioselectivity can be explained in terms of facilitation of deprotonation of **N** by the electron-withdrawing carbonyl group (i.e., kinetic acidity control).

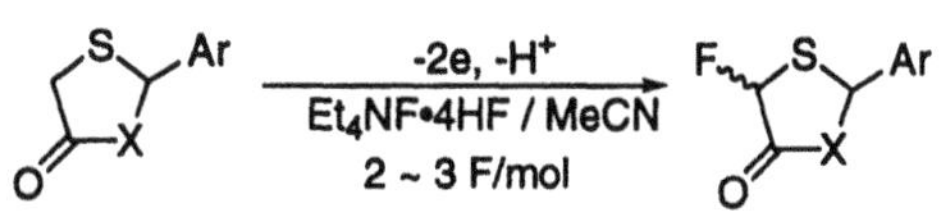

51: X= O
52: X= S

X= O, Ar = Et: 86% (trans/cis: 53/47)
Ar = Ph: 70% (trans/cis: 55/45)
X= S, Ar = n-Pr: 78% (trans/cis: 52/48)
Ar = Mesityl: 58% (trans/cis: 80/20)

Scheme 39.

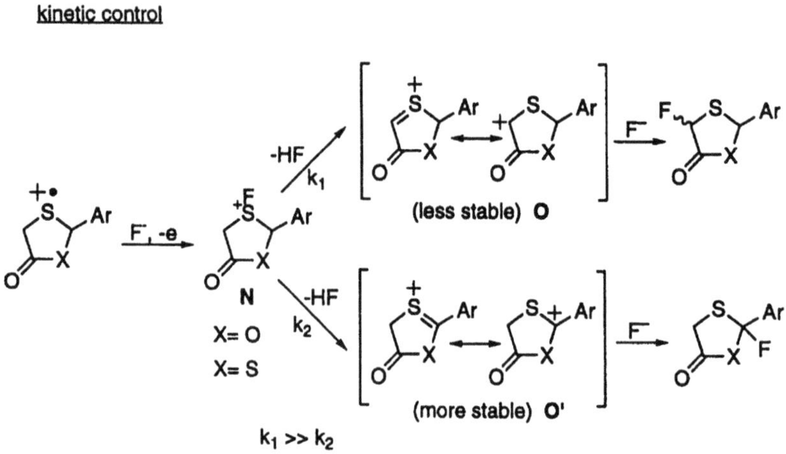

Scheme 40.

Therefore, such high regioselectivity as is observed in these anodic fluorinations is reasonable.

Highly regioselective anodic monofluorination of 2-aryl-4-thiazolidinones **53** can also be promoted by using pulse electrolysis in $Et_3N{\cdot}3HF$/MeCN as shown in Scheme 41.[185] In this case, benzylic fluorination does not take place and fluorination occurs at the 5-position selectively. The regioselectivity of this process appears also to be controlled kinetically as shown in Scheme 40. The transformation of the sulfones **55** derived from fluorinated products **54** into monofluoro β-lactams **56** is readily performed in excellent yields by thermolysis (Scheme 41).

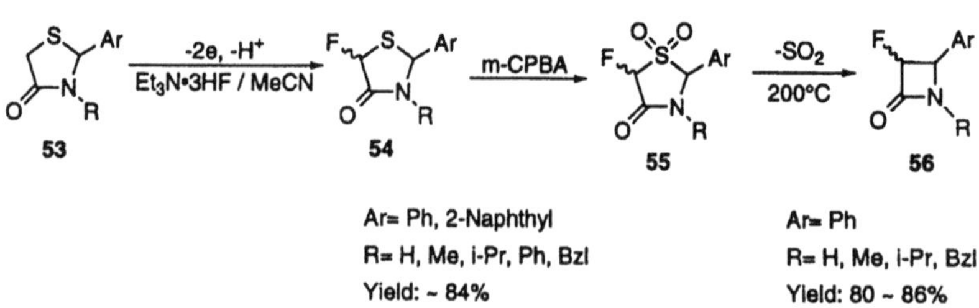

Scheme 41.

N-Substituted 2*H*-1,4-benzothiazin-3(4*H*)-one derivatives **57** undergo selective and efficient anodic fluorination in $Et_3N{\cdot}3HF$/MeCN to give the corresponding fluorinated products in good yields (Eq. 110).[186] In these cases, fluorination does not take place on the benzene ring. When R in **57** is hydrogen, only polymerized products are formed.

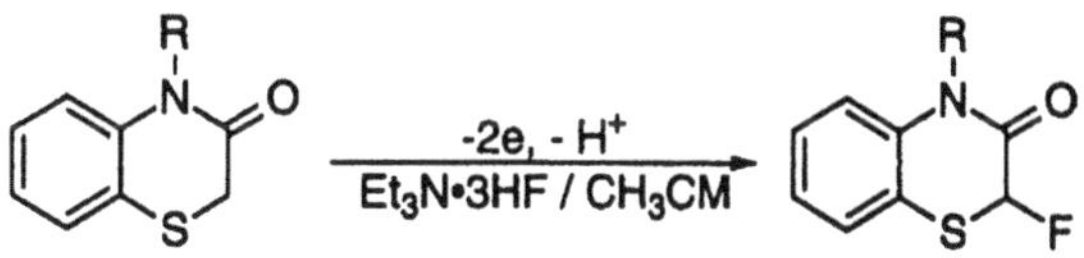

57 (110)

R = Me: 68%, i-Pr: 88%, COPh: 77%

In contrast to 2*H*-1,4-benzothiazin-3(4*H*)-ones **57**, anodic oxidation of 3*H*-1,4-benzoxathian-2-ones **58** in $Et_3N{\cdot}3HF$ gives poor yields of the desired fluorinated products **59** owing to strong passivation of the anode. However, use of $Et_4NF{\cdot}m$HF ($m = 3$ or 4)/MeCN suppresses passivation and leads to good to excellent yields of **59** on pulse electrolysis.[187] In particular, when $Et_3NF{\cdot}4HF$ is used as a supporting electrolyte and the fluoride source, electrolysis can be carried out with a large current density. Therefore, the anodic fluorination is completed in a short time period and excellent product yields were obtained as shown in Table 18. The increases observed in the yields and current

Table 18. Anodic Monofluorination of 3*H*-1,4-Benzoxathian-2-ones

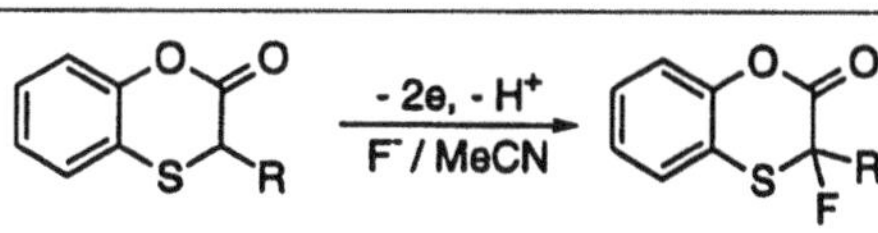

58 **59**

Run	*R*	*Anodic Potential*	*Supporting Electrolyte*	*Electricity (F/mol)*	*Time (h)*	*Yield (%)*
1	H (**58a**)	1.8[a]	$Et_3N{\cdot}3HF$	2.1	18.5	5
2	H (**58a**)	1.8[a]	$Et_4NF{\cdot}3HF$	2.2	9	54
3	H (**58a**)	1.8[b]	$Et_4NF{\cdot}4HF$	2.4	3	80
4	Me (**58b**)	1.8[b]	$Et_4NF{\cdot}4HF$	2.4	3.5	82

Notes: [a] V versus SSCE.
[b]V versus SSCE.

efficiencies are mainly attributable to the much higher oxidation potential of $Et_4NF \cdot 4HF$/MeCN compared with that of $Et_3N \cdot 3HF$/MeCN.

Various lactams have also been anodically fluorinated efficiently. Although five- and six-membered lactams were selectively monofluorinated, seven-membered lactams did not give desired fluorinated products (Scheme 42).[186]

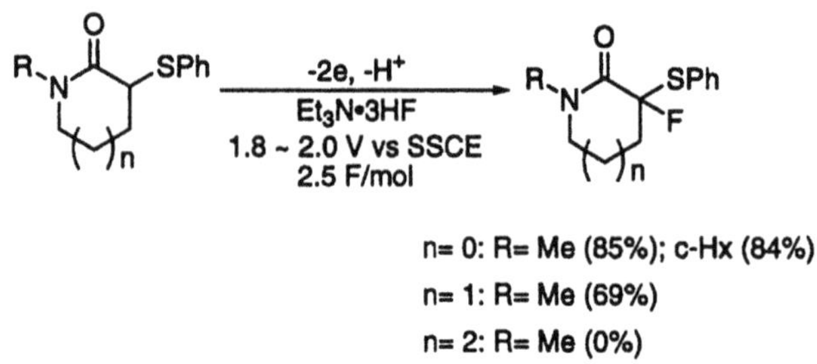

Scheme 42.

Anodic monofluorination of α-phenylthio-β-lactams **60** in $Et_3N \cdot 3HF$/MeCN also leads to production of the corresponding α-fluorinated products in good yields and with high current efficiencies as shown in Eq. 111.[188]

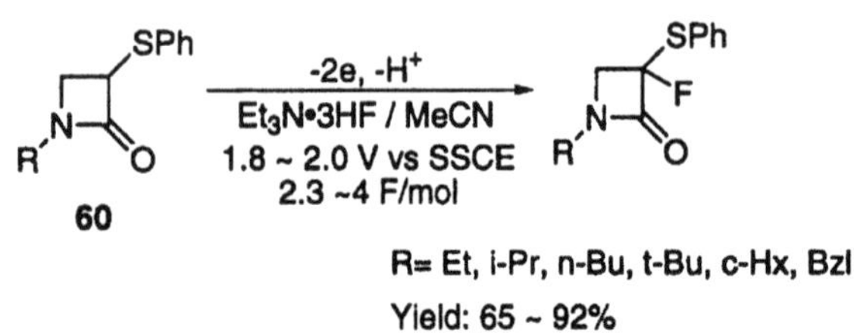

60

(111)

Suda and his co-workers reported anodic fluoro desilylation of β-trimethylsilyl-β-lactams as shown in Scheme 43.[189] In this reaction, the selection of the trimethylsilyl as a good leaving group enables regioselectivity control in the fluorination process.

Monofluoro β-lactams are used not only as synthetic intermediates for the preparation of fluorinated β-lactam antibiotics but also as precursors of carbohydrates and amino acids. However, there are a limited number of reports on the synthesis of such compounds and

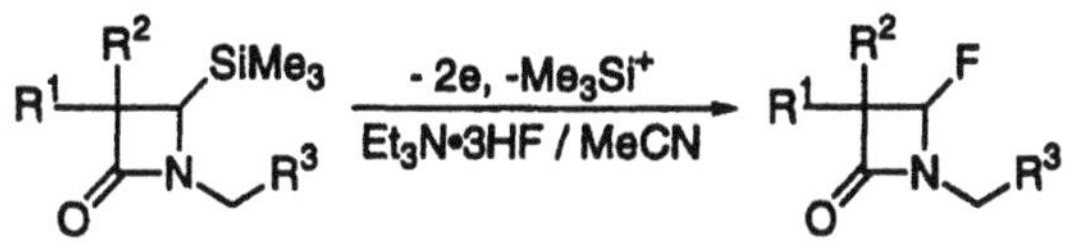

$R^1 = R^2 = H, R^3 = Ph$: 80%, Pr : 78%
$R^1 = H, R^2 = MeCH, R^3 = Ph$: 80%
$R^1 = H, R^2 = CH_3CH_2O(Me)CH, R^3 = Ph$: 75%

Scheme 43.

only two examples of direct fluorination of β-lactam derivatives have been described to date. In both cases, the dangerous substance $FClO_4$ is used as the fluorinating reagent. Therefore, anodic monofluorination of β-lactams has greater preparative applicability than conventional chemical methods.

Anodic monofluorination of the biologically interesting oxindole **61** has been described.[190,191] This process was greatly affected by use of supporting fluoride salts. $Et_4NF{\cdot}4HF$ was the most effective salt for this purpose (Table 19) when used with a carbon or platinum anode. The 3-oxo-1,2,3,4-tetrahydroisoquinoline derivative **63** was success-

Table 19. Anodic Monofluorination of 3-(Phenylthio)oxindole Derivatives

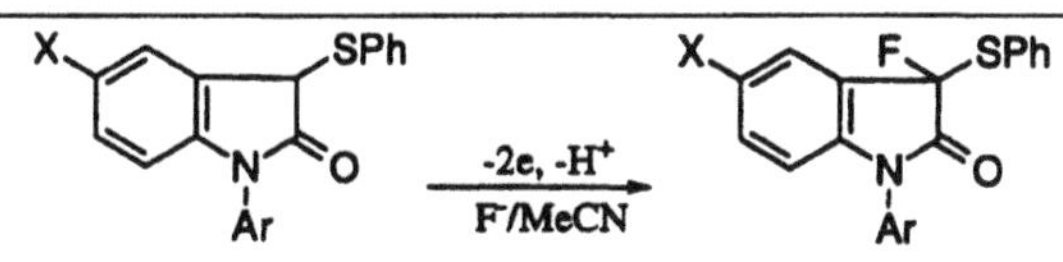

61a: Ar = Ph; X = H
61b: Ar = *p*-Tol; X = Me

62a: Ar = Ph; X = H
62b: Ar = *p*-Tol; X = Me

Oxindole	*Supporting Electrode*	*Charge Passed (F/mol)*	*Yield of* **2** *(%)*
61a	$Et_3N{\cdot}3HF$	6	**62a** 30
61a	$Et_4NF{\cdot}2HF$	4	**62a** 41
61a	$Et_4NF{\cdot}3HF$	3.7	**62a** 58
61a	$Me_4NF{\cdot}4HF$	3.5	**62a** 64
61b	$Et_4NF{\cdot}3HF$	3	**62b** 20
61b	$Me_4NF{\cdot}4HF$	3	**62b** 50

fully fluorinated under these conditions (Eq. 112).[191] Although **63** has three kinds of benzylic carbons, fluorination takes place at the 4-position exclusively. In contrast to **61** and **63**, corresponding oxindoles and 3-oxo-1,2,3,4-tetrahydroisoquinoline derivatives which are devoid of a phenylthio group do not undergo selective anodic fluorination, but rather yield a complicated mixture of products as a result of competitive fluorination of the benzene rings. Therefore, the phenylthio group is essential to promote selective fluorination in these systems. Consequently, the initial step of this anodic fluorination should be electron transfer from the sulfur atom in the molecules. In contrast, Laurent and his co-workers reported that anodic fluorination of thiofluvone provided mono- and/or trifluorinated products depending on the applied potential as shown in Eq. 113.[192]

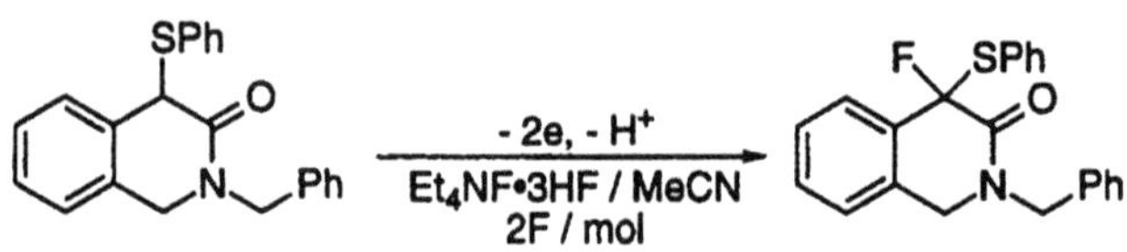

63

(112)

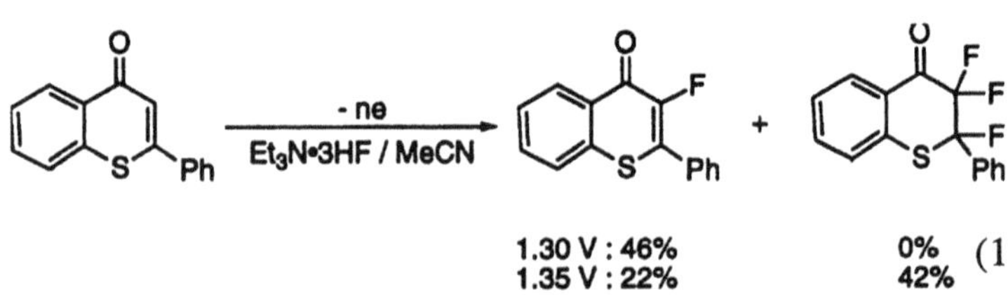

(113)

Recently, Yoshida and his co-workers found an example of an anodic process involving direct displacement of a tributylstannyl group with fluorine. They extended this fluorination process to accomplish unique cyclization–fluorination reactions as shown in Scheme 44.[193] Similar fluorocyclization methods based on the use of an organothio group instead of the stannyl group have been developed (Scheme 45).[194]

Anodic Monofluorination of the Side Chain of Heterocycles

Fuchigami and his co-workers have systematically studied anodic monofluorination reactions which occur at the side chain of various

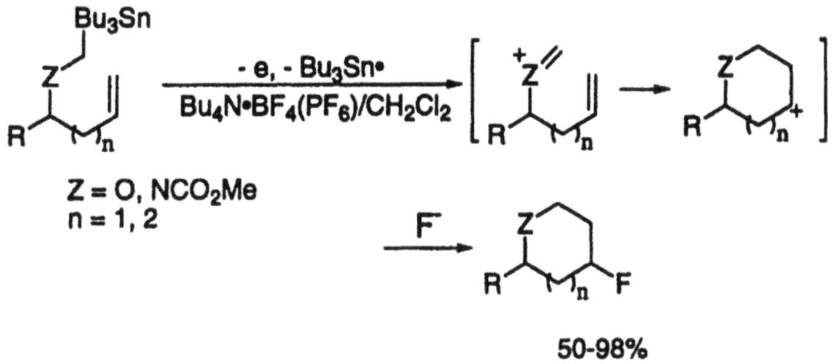

Scheme 44.

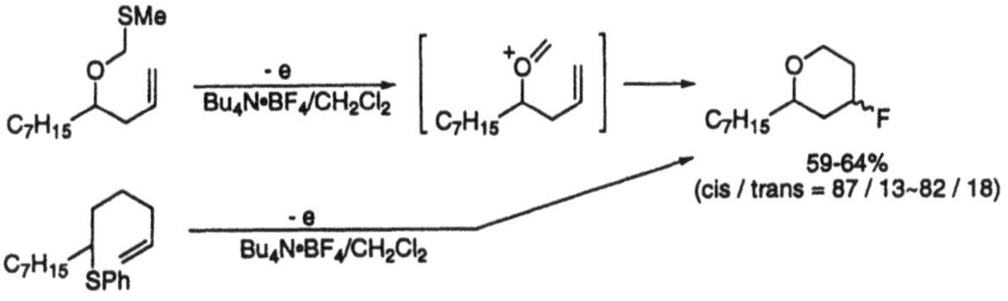

Scheme 45.

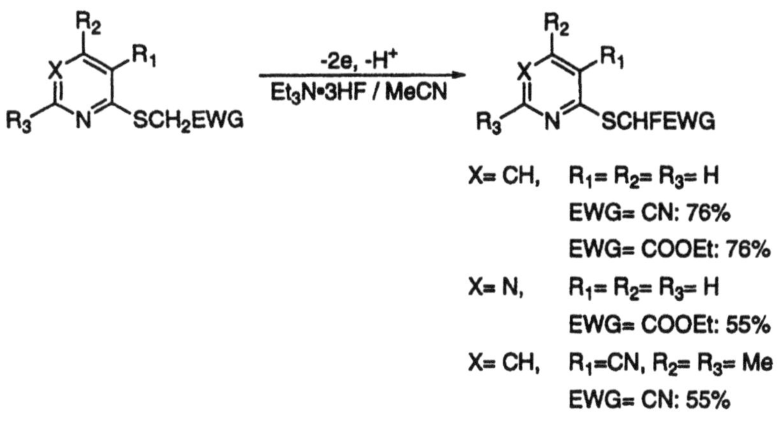

Scheme 46.

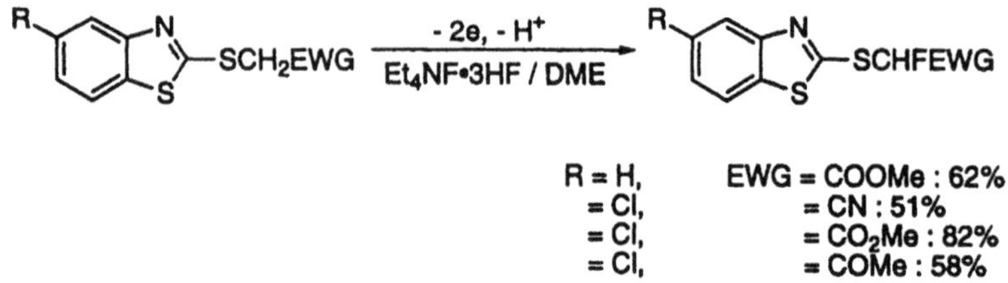

Scheme 47.

heterocyclic compounds. As shown in Schemes 46 and 47, the active methylenethio group attached to heterocycles is selectively fluorinated to give the corresponding α-fluorinated products in good yields.[195–197] In these reactions, the choice of solvent and supporting fluoride salt is of great importance in order to effect a successful fluorination reaction.[197] The fluorinated 2-pyridyl sulfides produced in this manner are easily converted into 2-fluorothienio[2,3-*b*]pyridines in good yields (Scheme 48).[196]

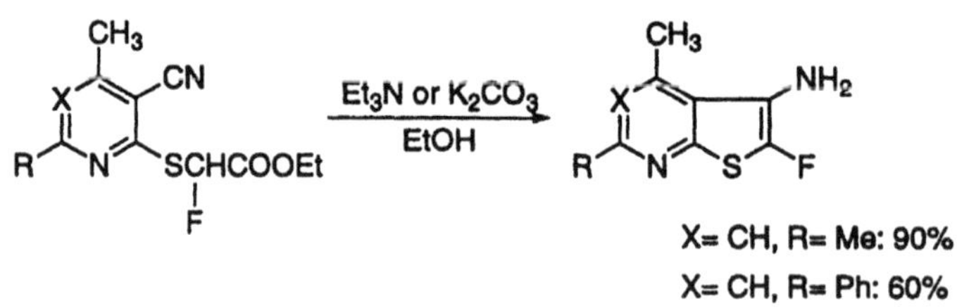

Scheme 48.

10.9. Anodic *gem*-Difluorination

Organic compounds containing a difluoromethylene group are of great interest in biochemistry as this functionality is isopolar and isosteric with an ether oxygen (Figure 3). The difluoromethylene group is typically prepared by using various reagents such as molybdenum hexafluoride, selenium tetrafluoride, sulfur tetrafluoride, and (dimethylamino)sulfur trifluoride (DAST). However, these reagents are highly toxic and their use requires severe reaction conditions. In contrast to the chemical methods, anodic difluorinations can be carried out under safe conditions without the use of oxidizing reagents. For example, anodic fluorination of monofluoro α-thio-substituted ester

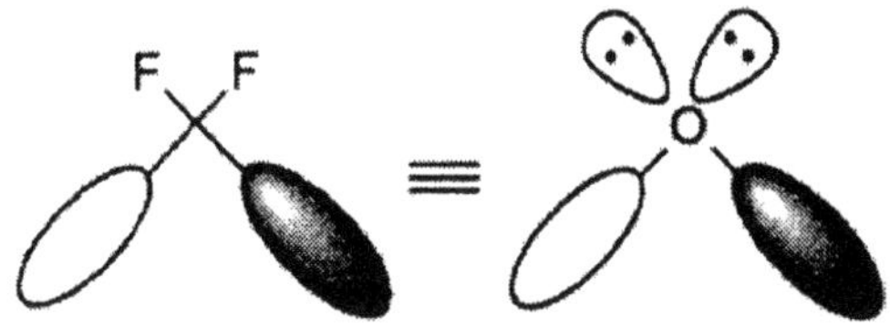

Figure 3.

41a provides two *gem*-difluoro products depending on the work-up procedure used. When the electrolytic solution is neutralized with either aqueous $NaHCO_3$ or aqueous ammonia, the *gem*-difluoro ester **64a** or amide **64b** is formed as shown in Scheme 49.[161]

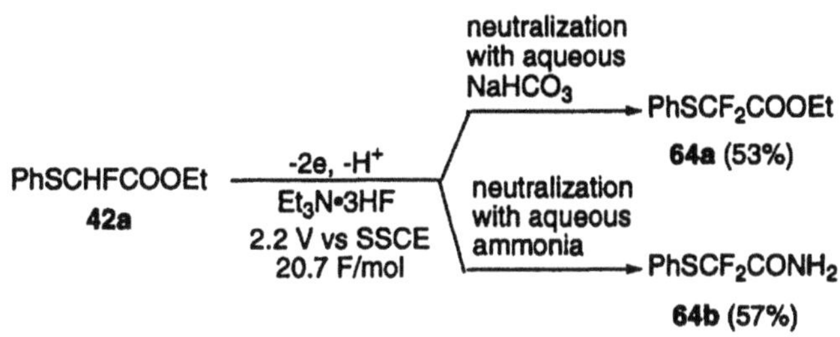

Scheme 49.

However, the current efficiency for this difluorination reaction is extremely low because of competitive oxidation of $Et_3N{\cdot}3HF$ and the starting compound. On the other hand, *gem*-difluorinated products **64** can be obtained directly from the organic sulfur compounds **41** by using the less oxidizable salt $Et_4NF{\cdot}4HF$ as shown in Scheme 50.[198] In these cases, theoretical amounts of electricity must be employed to bring about complete fluorination.

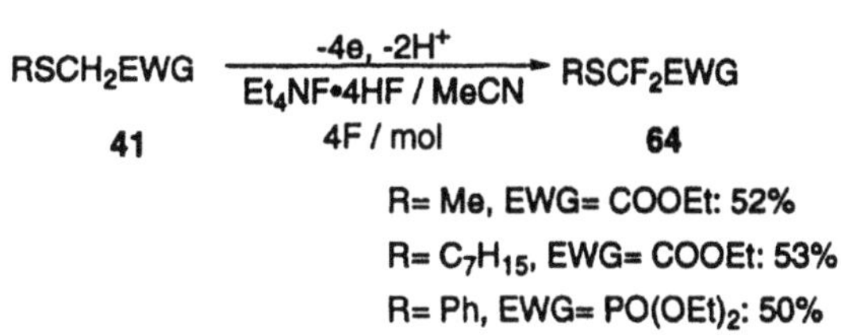

Scheme 50.

Recently, oxidative fluorodesulfurization of dithioacetals such as 1,3-dithiolanes and 1,3-dithianes using *N*-halo compounds and pyridinium poly(hydrogen fluoride), $Bu_4N{\cdot}H_2F_3$, or 4-methyl(difluoroiodo)benzene has been employed as a synthetic method. In these reactions, large amounts of the *N*-halo reagents or of 4-methyl(difluoroiodo)benzene are required. It was expected that the electrochemical oxidation technique could be applied to these fluorodesulfurization processes. In fact, anodic desulfurization of dithioacetals **65** in the presence of $Et_3N{\cdot}3HF$ takes place to provide the corresponding *gem*-difluoro compounds **66** while dithioacetals of aromatic and aliphatic aldehydes (e.g., **67** and **70**) react to give *gem*-difluoro thioethers **69** and monofluoro thioethers **71**, respectively, as shown in Eqs. 114–116.[199]

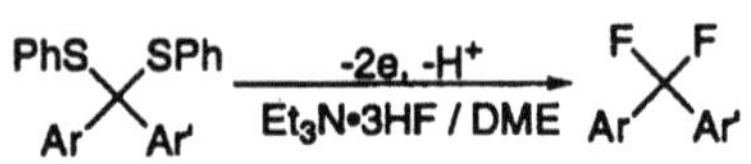

(114)

65 **66**

Ar= Ar'= Ph: 53%
Ar= Ar'= *p*-FC_6H_4: 79%
Ar= Ph, Ar'= *p*-FC_6H_4: 70%

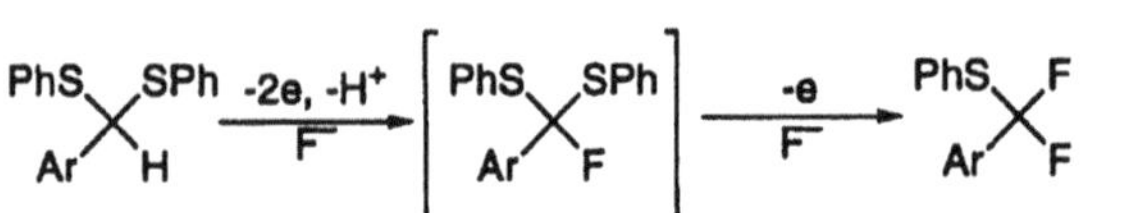

(115)

67 **68** **69**

(Ar= *p*-$NO_2C_6H_4$: 57%)

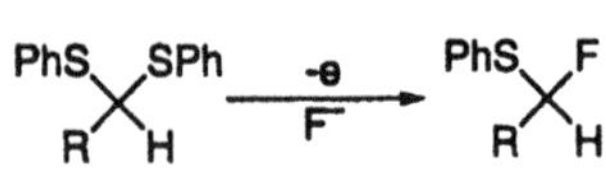

(R= C_9H_{19}: 54%) (116)

70 **71**

It is quite interesting that the aromatic aldehyde derived dithioacetal **67** shows different anodic behavior from that of the aliphatic aldehyde derived dithioacetal **70**. This can be explained by the pathway shown in Scheme 51. Deprotonation of the cation radical **P** arising from the aromatic dithioacetal **67** should be more facile than that from **P′** which comes from the alkyl dithioacetal **70** as the former contains more acidic α-hydrogens.

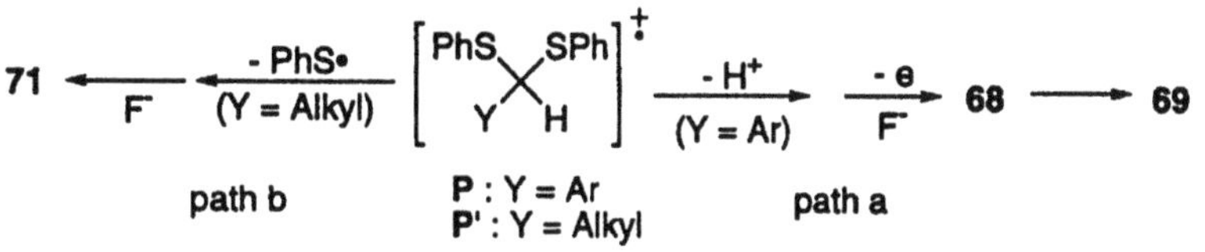

Scheme 51.

Hydrazones are known to be alternative precursors to *gem*-difluoromethylene compounds. Oxidative *gem*-difluorination of hydrazones has been performed with fluorine gas, *in situ* prepared IF from I_2 and F_2, or *in situ* generated BrF from *N*-bromosuccinimide/Py$(HF)_n$. In contrast, anodic oxidation of hydrazones in the presence of fluoride ions provides monofluorinated products predominantly along with smaller amounts of *gem*-difluorinated products (Scheme 52).[200]

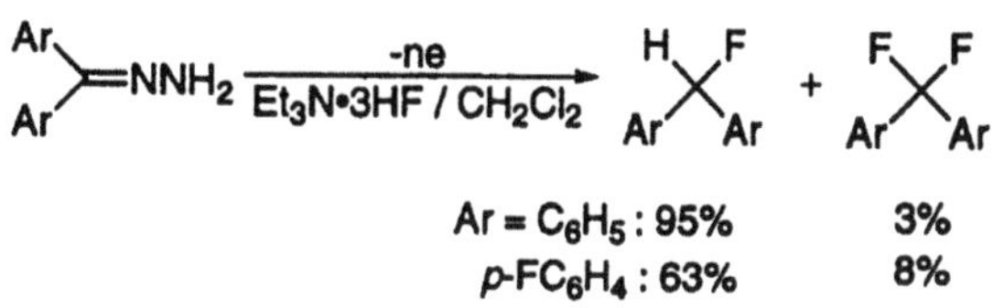

Scheme 52.

10.10. Indirect Anodic Fluorination

As mentioned above, passivation of the electrode as a result of the formation of nonconductive polymers takes place commonly during anodic oxidation of organic substrates in the presence of fluoride ions. For example (Eq. 114), anodic oxidative difluorodesulfurization of dithioacetals does occur, although the current efficiencies are low because of this passivation phenomenon.[199] In order to avoid this complication, Fuchigami and his co-workers have developed an indirect electrochemical method in which various mediators are used. The indirect anodic difluorodesulfurization of dithioacetals was performed by using a Br^+/Br^- redox mediator.[201] A cyclic voltammetric study of Br^- in $Et_3N{\cdot}3HF/CH_2Cl_2$ suggested that mixed polyhalide ions such as Br_2F^- and BrF_2^- are generated at the anode under these conditions. The mediator was found to be effective for oxidative

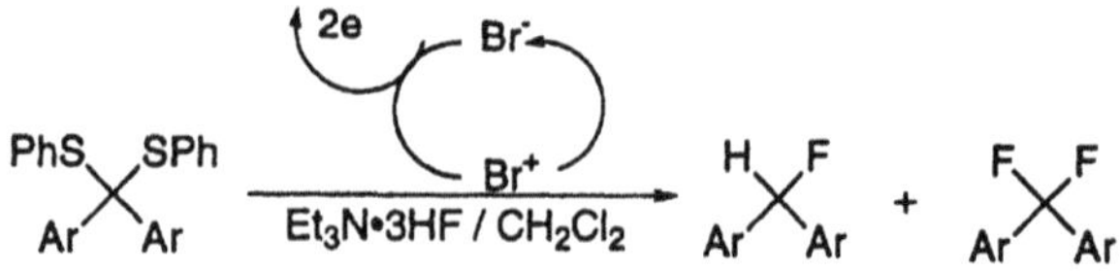

1.5 F/mol :	Ar = Ph :	93	0
4.5 F/mol		trace	32
1.5 F/mol :	Ar = *p*-FC_6H_4 :	quant	0
3.7 F/mol :		27	43

Scheme 53.

monofluorodesulfurization of dithioacetals but the efficiency for difluorodesulfurization was low (Scheme 53). This is quite different from the chemical process using *N*-bromo reagents as the oxidizing agents. In the redox-mediated system, no electrode passivation was observed during the fluorination. This appears to be the first example of a successful indirect electrochemical monofluorination reaction.[201]

Although hypervalent iodobenzene difluorides are selective fluorinating reagents, their preparation is not straightforward and these reagents are not stable. Fuchigami and his co-workers have found that anodic oxidation of *para*-substituted iodobenzenes in the presence of fluoride ions provides the corresponding iodobenzene difluorides in high yields.[202] These workers also demonstrated that selective indirect anodic desulfurizative coupled with *gem*-difluorination can be successfully carried out using a catalytic amount of iodobenzene difluoride **72** as shown in Scheme 54. This is the first example of successful indirect anodic *gem*-difluorination.

Direct anodic oxidation of iodobenzene and *para*-methyliodobenzene did not give the corresponding hypervalent difluorides. Therefore, indirect anodic oxidation was attempted using a chloride ion mediator.[203] In this case, the desired difluoro products are not formed but instead a novel hypervalent iodobenzene derivative, such as **73**, having I–F and I–Cl bonds is produced. It was found that difluorination occurs by the reaction of dithioacetals with iodobenzene chlorofluorides. Indirect anodic oxidation of dithioacetals by using a catalytic amount of *para*-methoxyiodobenzene was attempted as shown in Scheme 55 and it was found that desulfurizative difluorina-

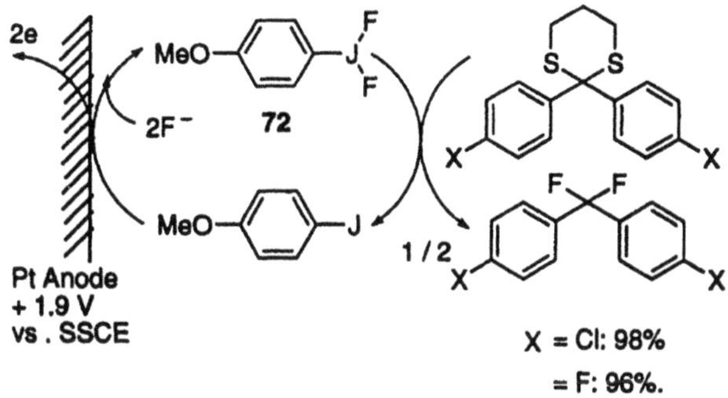

Scheme 54.

tion proceeds smoothly at a low oxidation potential of 1.3 V versus SCE, where only chloride ions are discharged, to provide *gem*-difluoro products in good yields.[203]

Radical cations of triarylamines are known to be single electron transfer oxidizing agents. Very recently, these substances have been used as efficient mediators for *gem*-difluorodesulfurization as shown in Scheme 56.[204] Furthermore, triarylamines have recently been shown to be highly effective mediators for monofluorodesulfurization of β-lactams.[205] Severe passivation of the anode takes place during anodic fluorination of β-phenylthio- β-lactams, which is quite different from the case of α-phenylthio-β-lactams **60**. However, selective anodic fluorodesulfurization proceeds efficiently without passivation when tris(2,4-dibromophenyl)amine is employed as a mediator as shown in Scheme 57.[205]

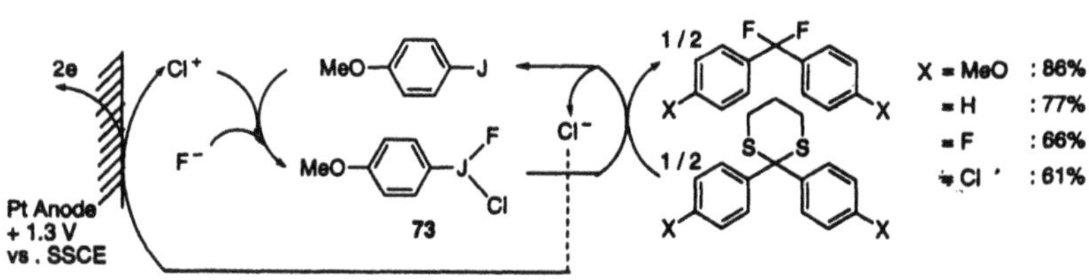

Scheme 55.

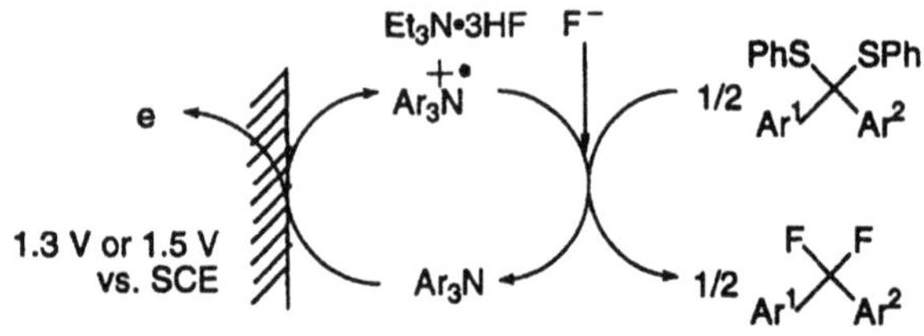

Ar = 4-BrC_6H_4(1.3V), Ar^1 = *p*-FC_6H_4, Ar^2 = Ph : 83%
Ar = 3,4-$Br_2C_6H_3$(1.5V), Ar^1 = Ar^2 = *p*-ClC_6H_4 : 74%

Scheme 56.

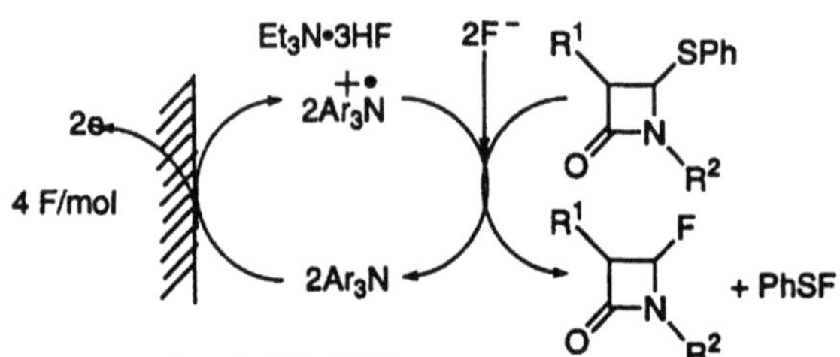

Ar = 2,4-$Br_2C_6H_3$

R^1= H; R^2= $C_6H_5CH_2$: 83%
R^1= H; R^2= *p*-$BrC_6H_4CH_2$: 100%
R' = Me_2SiOCH–; R^2 = $C_6H_5CH_2$: 66%
t-Bu Me

Scheme 57.

10.11. Chemical Fluorination Using Fluorinating Reagents

Chemical fluorination of the types of compounds mentioned above was also examined using commercially available fluorinating reagents such as *N*-fluoropyridinium salts and Et_2NSF_3 (DAST) as these reagents are known to be effective for fluorinations of sulfides.[206–208] However, the use of these reagents did not give rise to formation of any of the desired fluorination products in most cases or to only low yields of these materials. Therefore, electrochemical fluorination is superior to the conventional chemical fluorination methodologies.

11. CONCLUSIONS

Organic electrochemistry is an inherently hybrid and interdisciplinary research field. Despite the recent increased importance of organofluorine compounds, the electrochemical electron transfer reactions did not become recognized as a powerful tool for organofluorine synthesis until about 10 years ago. Great progress has been made in this area and various new methodologies have been developed for the preparation of organofluorine compounds. It is hoped that this chapter has demonstrated that electrochemical methodology represents a novel and valuable tool for both the selective fluorination of organic molecules and molecular conversion of fluoro-organic compounds. Hopefully, this review will stimulate organic chemists to further develop the highly valuable field of organofluorine electrochemistry.

ACKNOWLEDGMENTS

I wish to express my sincere thanks to my collaborators who are cited in the references. I am also grateful to Mr. Yankun Hou, a graduate student in my research group, for his assistance in preparing this chapter.

REFERENCES

1. Knunyants, I. L.; Yakobson, G. G., Eds. *Synthesis of Fluoroorganic Compounds*; Springer-Verlag: Berlin, 1985.
2. Welch, J. T., Ed. *Selective Fluorination in Organic and Bioorganic Chemistry*; Am. Chem. Soc.: Washington, DC, 1991.
3. Welch, J. T.; Eswarakrishnan, S., Eds. *Fluorine in Bioorganic Chemistry*; Wiley: New York, 1991.
4. Olah, G. A.; Chambers, R. D.; Prakash, G. K. S., Eds. *Synthetic Fluorine Chemistry*; Wiley: New York, 1992.
5. Hudlicky, M.; Pavlath, A. E., Eds. *Chemistry of Organic Fluorine Compounds II. A Critical Review*; Am. Chem. Soc.: Washington, DC, 1995.
6. Begue, J.-P.; Bonnet-Delpon, D. *Tetrahedron* **1991**, *47*, 3207.
7. McClinton, M. A.; McClinton, D. A. *Tetrahedron* **1992**, *32*, 6555.
8. Tozer, M. J.; Herpin, T. F. *Tetrahedron* **1996**, *52*, 8619.
9. Chambers, R. D., Ed. *Topics in Organofluorine Chemistry*; Springer-Verlag: Berlin, 1997.
10. Fuchigami, T. *Top. Curr. Chem.* **1994**, *170*, 1.
11. Inesi, A.; Rampazzo, N. J. *J. Electroanal. Chem.* **1974**, *49*, 85.
12. Kopilov, J.; Evans, D. H. *J. Electroanal. Chem.* **1990**, *280*, 435.

13. Coleman, J. P.; Naser-ud-din; Gilde, H. G.; Utley, H. P.; Weedon, C. L.; Eberson, L. *J. Chem. Soc. Perkin Trans. II* **1973**, 1903.
14. Uneyama, K.; Maeda, K.; Kato, T.; Katagiri, T. In *Novel Trends in Electroorganic Synthesis*; Torii, S., Ed.; Springer: Berlin, 1998, p. 301.
15. Barker, D. J.; Brewis, D. M.; Dahn, R. H. *Electrochim. Acta* **1978**, *23*, 1107.
16. Marsella, J. A.; Gilicinslci, A. G.; Coughlin, A.; Pez, P. G. *J. Org. Chem.* **1992**, *57*, 2856.
17. Kariv-Miller, E.; Vaitner, Z. *J. Org. Chem.* **1985**, *50*, 1394.
18. Rozhkov, I. N.; Igumnov, S. N.; Pletnev, S. I.; Borisov, Y. A.; Rempel, G. D. In *6th Japanese–Soviet Symposium on Fluorine Chemistry*, Novosibirsk, USSR, 11–13 July 1989.
19. Andrieux, C. P.; Gelis, L.; Medebielle, M.; Pinson, J.; Saveant, J. M. *J. Am. Chem. Soc.* **1990**, *112*, 3509.
20. Calas, P.; Moreau, P.; Commeyras, A. *J. Electroanal. Chem.* **1977**, *78*, 271.
21. Ignat'ev, N. V.; Datsenko, S. D.; Yagupolskii, L. M. *Zh. Org. Khim.* **1990**, *26*, 1740.
22. Ignat'ev, N. V.; Datsenko, S. D.; Yagupolskii, L. M. *Zh. Org. Khim.* **1991**, *27*, 905.
23. Calas, P.; Moreau, P.; Commeyras, A. *J. Fluorine Chem.* **1978**, *12*, 67.
24. Calas, P.; Moreau, P.; Commeyras, A. *J. Chem. Soc. Chem. Commun.* **1982**, 433.
25. Shono, T.; Kise, N.; Oka, H. *Tetrahedron Lett.* **1991**, *32*, 6567.
26. Medaghri-Alaoui, A.; Choukroun, H.; Calas, P.; Amatore, C.; Commeyras, A. *J. Fluorine Chem.* **1991**, *54*, 71.
27. Smertenko, E. A.; Datsenko, S. D.; Ignat'ev, N, V.; Deev, L. E.; Bil'dinov, I. K.; Podsevalov, P. V. *Russ. J. Electrochem.* **1994**, *30*, 1172.
28. Andrieux, C. P.; Gelis, L.; Saveant, J. M. *Tetrahedron Lett.* **1989**, *30*, 4961.
29. Andrieux, C. P.; Gelis, L.; Saveant, J. M. *J. Am. Chem. Soc.* **1990**, *112*, 786.
30. Medebielle, M.; Pinson, J.; Saveant, J. M. *Tetrahedron Lett.* **1990**, *31*, 1279.
31. Medebielle, M.; Pinson, J.; Saveant, J. M. *J. Am. Chem. Soc.* **1991**, *113*, 6872.
32. Medebielle, M.; Oturan, M. A.; Pinson, J.; Saveant, J. M. *J. Org. Chem.* **1996**, *61*, 1331.
33. Medebielle, M. *Tetrahedron Lett.* **1995**, *36*, 2071.
34. Ignat'ev, N. V.; Datsenko, S. D.; Yagupolskii, L. M. *Sov. Electrochem.* **1991**, *27*, 1016.
35. Ignat'ev, N. V.; Datsenko, S. D.; Yagupolskii, L. M. *Sov. Electrochem.* **1992**, *28*, 405.
36. Sakai, T.; Yan, F.; Kashino, S.; Uneyama, K. *Tetrahedron* **1995**, *52*, 233.
37. Uneyama, K.; Yanagiguchi, T.; Asai, H. *Tetrahedron Lett.* **1997**, *38*, 7763.
38. Chaussard, J.; Folest, J. C.; Nédélec, J. Y.; Périchon, F.; Sibille, S.; Troupel, M. *Synthesis* **1990**, 369.
39. Sibille, S.; d'Incan, E.; Leport, L.; Périchon, J. *Tetrahedron Lett.* **1986**, *27*, 3129.
40. Sibille, S.; Mcharek, S.; Périchon, J. *Tetrahedron* **1989**, *27*, 3129.

41. Sibille, S.; Périchon, J.; Chaussard, J. *Synth. Commun.* **1989,** *27,* 3129.
42. Saboureau, C.; Trouple, M.; Sibille, S.; Périchon, J. *J. Chem. Soc. Chem. Commun.* **1989,** 1138.
43. Bordeau, M.; Brian, C.; Serein-Spirau, F.; Leger-Lambert, M.-P.; Deffievx, D.; Moreau, C.; Dunogues, J. In *Electrochemistry in the Preparation of Fluorine and Its Compounds*; Childs, W. V.; Fuchigami, T., Eds.; Electrochem. Soc.: Pennington, 1997, p. 201.
44. Prakash, G. K. S.; Deffievx, D.; Yudin, A. K.; Olah, G. A. *Synlett* **1994,** 1057.
45. Folest, J. C.; Nédélec, J. Y.; Périchon, J. *Synth. Commun.* **1988,** *18,* 1491.
46. Nédeélec, J. Y.; Mouloud, H. A. H.; Folest, J. C.; Périchon. J. *J. Org. Chem.* **1988,** *53,* 4720.
47. Paratian, J. M.; Sibille, S.; Périchon, J. *J. Chem. Soc. Chem. Commun.* **1992,** 53.
48. Mcharek, S.; Sibille, S.; Nédélec, J. Y.; Périchon, J. *J. Organomet. Chem.* **1991,** *401,* 211.
49. Meyer, G.; Rollin, Y.; Périchon, J. *J. Organomet. Chem.* **1987,** *333,* 263.
50. Rollin, Y.; Trouple, M.; Talec, D. G.; Périchon, J. *J. Organomet. Chem.* **1986,** *303,* 131.
51. Saboureau, C.; Sibille, S.; Troupel, M. *J. Chem. Soc. Chem. Commun.* **1989,** 885.
52. Sibille, S.; Ratovelomanana, V.; Périchon, J. *J. Chem. Soc. Chem. Commun.* **1992,** 283.
53. Uno, H.; Ochi, M.; Kunugi, A. In *Electrochemistry in the Preparation of Fluorine and Its Compounds*; Childs, W. V.; Fuchigami, T., Eds.; Electrochem. Soc.: Pennington, 1997, p. 190.
54. Iwasaki, T.; Yoshiyama, A.; Sato, N.; Fuchigami, T.; Nonaka, T.; Sasaki, M. *J. Electroanal. Chem.* **1987,** *238,* 315.
55. Iwasaki, T.; Sasaki, M.; Sato, N.; Yoshiyama, T.; Nonaka, T. *Denki Kagaku* **1990,** *58,* 83.
56. Briscoe, M. W.; Chambers, R. D.; Silvester, M. J.; Drakesmith, F. G. *Tetrahedron Lett.* **1988,** *29,* 1295.
57. Fuchigami, T.; Nakagawa, Y. *J. Org. Chem.* **1987,** *52,* 5276.
58. Shono, T.; Ishifune, M.; Okada, T.; Kashimura, S. *J. Org. Chem.* **1991,** *56,* 2.
59. Barhdadi, R.; Gal, J.; Heintz, M.; Troupel, M. *J. Chem. Soc. Chem. Commun.* **1992,** 50.
60. Becker, J. Y.; Smart, B. E.; Fukunaga, T. *J. Org. Chem.* **1988,** *53,* 5714.
61. Germain, A.; Commeyras, A. *Tetrahedron* **1981,** *37,* 487.
62. Germain, A.; Brunel, D.; Moreau, P. *Bull. Soc. Chim. Fr.* **1986,** 895.
63. Germain, A.; Brunel, D.; Moreau. P. *J. Fluorine Chem.* **1989,** *43,* 249.
64. Brunel, D.; Germain, A.; Moreau, P. *J. Chem. Soc. Perkin Trans. I* **1989,** 2283.
65. Buchner, W.; Garreau, R.; Lemaire, M. *J. Electroanal. Chem.* **1990,** *277,* 355.
66. Cassidy, J.; Breen, W. *Synth. Met.* **1991,** *41–43,* 3059.
67. Fuchigami, T.; Nakagawa, Y.; Nonaka, T. *Tetrahedron Lett.* **1986,** *27,* 3869.
68. Fuchigami, T.; Nakagawa, Y.; Nonaka, T. *J. Org. Chem.* **1987,** *52,* 5489.

69. Fuchigami, T.; Yamamoto, K.; Konno, A. *Tetrahedron* **1991**, *47*, 625.
70. Fuchigami, T.; Yamamoto, K.; Nakagawa, Y. *J. Org. Chem.* **1991**, *56*, 137.
71. Fuchigami, T.; Ichikawa, S.; Konno, A. *Chem. Lett.* **1989**, 1987.
72. Fuchigami, T.; Ichikawa, S.; Kandeel, Z. E.; Konno, A.; Nonaka, T. *Heterocycles* **1990**, *31*, 415.
73. Surowiec, K.; Fuchigami, T. *J. Org. Chem.* **1992**, *57*, 5781.
74. Fuchigami, T.; Ichikawa, S. *J. Org. Chem.* **1994**, *59*, 607.
75. Koizumi, T.; Fuchigami, T.; Nonaka, T. *Bull. Chem. Soc. Jpn.* **1989**, *62*, 219.
76. Fuchigami, T.; Yamamoto, K. *Chem. Lett.* **1996**, 937.
77. Koizumi, T.; Fuchigami, T.; Nonaka, T. *Chem. Lett.* **1987**, 1095.
78. Yoshida, J.; Isoe, S. *Chem. Lett.* **1987**, 631.
79. Kimura, M.; Koie, K.; Matsubara, S.; Sawaki, Y.; Iwamura, H. *J. Chem. Soc. Chem. Commun.* **1987**, 122.
80. Fuchigami, T.; Yamamoto, K.; Yano, H. *J. Org. Chem.* **1992**, *57*, 2946.
81. Fuchigami, T.; Shimojo, M.; Konno, A.; Nakagawa, K. *J. Org. Chem.* **1990**, *55*, 6074.
82. Konno, A.; Nakagawa, K.; Fuchigami, T. *J. Chem. Soc. Chem. Commun.* **1991**, 1027.
83. Fuchigami, T.; Yano, H.; Konno, A. *J. Org. Chem.* **1991**, *56*, 6732.
84. Shono, T.; Matsumura, Y.; Hayashi, J.; Mizoguchi, M. *Tetrahedron Lett.* **1980**, *21*, 1867.
85. Furuta, S.; Saito, Y.; Fuchigami, T. *J. Fluorine Chem.* **1998**, *87*, 209.
86. Furuta, S.; Fuchigami, T. *Electrochim. Acta* **1988**, *43*, 3153.
87. Fuchigami, T.; Sano, M. Unpublished results.
88. Fuchigami, T.; Sano, M. Unpublished results.
89. Jouikov, V. *Tetrahedron Lett.* **1993**, *34*, 6034.
90. Surowise, K.; Fuchigami, T. *Tetrahedron Lett.* **1992**, *33*, 1065.
91. Surowice, K.; Fuchigami, T. Submitted to *J. Org. Chem.*
92. Fuchigami, T.; Fujita, T.; Konno, A. Unpublished results.
93. Fujita, T.; Fuchigami, T. *Denki Kagaku* **1994**, *62*, 1161.
94. Fuchigami, T.; Fujita, Y.; Nonaka, T. *J. Electroanal. Chem.* **1990**, *284*, 115.
95. Shono, T.; Matsumura, Y.; Inoue, K.; Ohmizu, H.; Kashimura, S. *J. Am. Chem. Soc.* **1982**, *104*, 5753.
96. Furuta, S.; Fuchigami, T. *Electrochim. Acta* **1998**, *43*, 3183.
97. Konno, A.; Fuchigami, T. *Chem. Lett.* **1992**, 2405.
98. Konno, A.; Fuchigami, T.; Fujita, Y.; Nonaka, T. *J. Org. Chem.* **1990**, *55*, 1952.
99. Matsumura, Y.; Tomita, T.; Sudoh, M.; Kise, N. *Tetrahedron Lett.* **1994**, *35*, 1271.
100. Uneyama, K.; Kobayashi, M. *J. Org. Chem.* **1994**, *59*, 3003.
101. Kobayashi, M.; Uneyama, K. *J. Org. Chem.* **1996**, *61*, 3902.
102. Kobayashi, M.; Sadamune, K.; Mizukami, H.; Uneyama, K. *J. Org. Chem.* **1994**, *59*, 1909.
103. Fuchigami, T.; Ichigawa, S.; Konno, A. *Chem. Lett.* **1992**, 2405.
104. Fuchigami, T.; Furuta, S. Unpublished results.

105. Schäfer, H. J. *Top. Curr. Chem.* **1990**, *152*, 91.
106. Hein, L.; Cech, D. *Z. Chem.* **1977**, *17*, 415.
107. Brookes, C.; Coe, P. L.; Owen, D. M.; Pedler, A. E.; Tatlow, J. C. *J. Chem. Soc. Chem. Commun.* **1974**, 323.
108. Muller, N. *J. Org. Chem.* **1983**, *48*, 1370.
109. Muller, N. *J. Org. Chem.* **1984**, *49*, 2826.
110. Muller, N. *J. Org. Chem.* **1984**, *49*, 4559.
111. Muller, N. *J. Org. Chem.* **1986**, *51*, 263.
112. Muller, N. *J. Fluorine Chem.* **1987**, *36*, 163.
113. Uneyama, K.; Nanbu, H. *J. Org. Chem.* **1988**, *53*, 4598.
114. Uneyama, K.; Morimoto, O.; Nanbu, H. *Tetrahedron Lett.* **1989**, *30*, 109.
115. Uneyama, K.; Makio, S.; Nanbu, H. *J. Org. Chem.* **1989**, *54*, 872.
116. Uneyama, K.; Watanabe, S. *J. Org. Chem.* **1990**, *55*, 3909.
117. Uneyama, K. *Tetrahedron* **1991**, *47*, 555.
118. Uneyama, K.; Ueda, K. *Chem. Lett.* **1988**, 853.
119. Dan-oh, Y.; Uneyama, K. *Bull. Chem. Soc. Jpn.* **1995**, *68*, 2993.
120. Renaud, R. N.; Champagne, P. J. *Can. J. Chem.* **1975**, *53*, 529.
121. Seebach, D.; Renaud, P. *Helv. Chim. Acta* **1985**, *68*, 2342.
122. Kubota, T.; Aoyagi, R.; Sando, H.; Kawanishi, M.; Tanaka, T. *Chem. Lett.* **1987**, 1435.
123. Kubota, T.; Ishii, T.; Minamikawa, H.; Yamaguchi, S.; Tanaka, T. *Chem. Lett.* **1988**, 1987.
124. Cramer, E.; Hembrock, A.; Schäfer, H. J. Unpublished results.
125. Smertenko, E. A.; Datsenko, S. D.; Ignat'ev, N. V. *Russ. J. Electrochem.* **1998**, *39*, 46.
126. Purrington, S. T.; Kagen, B. S. *Chem. Rev.* **1986**, *86*, 997.
127. Rozen, S. *Acc. Chem. Res.* **1996**, *29*, 243.
128. Simons, J. H. U.S. Patents 1949, 2490098 and 2490099; *Chem. Abstr.* **1950**, *44*, 7443.
129. Childs, W. V. In *Technique Electroorganic Synthesis* Vol. V, Part III; Weinberg, N. L., Ed.; Wiley: New York, 1982.
130. Rozhkov, I. N. In *Organic Electrochemistry*, 2nd ed.; Baizer, M. M.; Lund, H., Eds.; Dekker: New York, 1983, Chapter 24.
131. Childs, W. V.; Christensen, L.; Klink, F. W.; Kolpin, C. F. In *Organic Electrochemistry*, 3rd ed.; Lund, H.; Baizer, M. M., Eds.; Dekker: New York, 1991, Chapter 26.
132. Wilkinson, J. A. *Chem. Rev.* **1992**, *92*, 505.
133. Umemoto, T. *Chem. Rev.* **1996**, *96*, 1757.
134. Schmidt, H.; Schmidt, H. D. *Chem. Tech.* **1953**, *5*, 454.
135. Schmidt, H.; Schmidt, H. D. *J. Prakt. Chem.* **1955**, *2*, 105.
136. Schmidt, H.; Meinert, H. *Angew. Chem.* **1960**, *72*, 109.
137. Rozhkov, I. N. *Russ. Chem. Rev.* **1976**, *45*, 615.
138. Knunyants, I. L.; Rozhkov, I. N.; Bukhtiarov, A. V.; Go'ldin M. M.; Kudryavvtseu, R. V. *Izv. Akad. Nauk SSSR, Ser. Khim.* **1970**, 1207.

139. Rozhkov, I. N.; Bukhtiarov, A. V.; Kuleshova, N. D.; Kudryavvtseu, R. V. *Dokl. Akad. Nauk SSSR* **1970**, *193*, 1322.
140. Rozhkov, I. N.; Gambaryan, N. P.; Galpern, E. G. *Tetrahedron Lett.* **1976**, 4819.
141. (a) Ludman, C. L.; McCarron, E. M.; O'Malley, R. F. J. *Electrochem. Soc.* **1972**, *119*, 874. (b) Meurs, J. H. H.; Sopher, D. W.; Eilenberg, W. *Angew. Chem. Int. Ed. Engl.* **1989**, *28*, 927.
142. Momota, K.; Morita, M.; Matsuda, Y. *Denki Kagaku* **1992**, *60*, 1016.
143. Momota, K.; Morita, M.; Matsuda, Y. *Electrochim. Acta*, **1993**, *38*, 619.
144. Momota, K.; Morita, M.; Matsuda, Y. *Electrochim. Acta* **1993**, *38*, 1123.
145. Bensadat, A.; Bodennec, G.; Laurent, E.; Tardivel, R. *Tetrahedron Lett.* **1977**, *18*, 3799.
146. Bensadat, A.; Bodennec, G.; Laurent, E.; Tardivel, R. *Nouv. J. Chim.* **1980**, *4*, 453.
147. Bensadat, A.; Bodennec, G.; Laurent, E.; Tardivel, R. *J. Fluorine Chem.* **1982**, *20*, 333.
148. Rozhkov, I. N.; Knunyants, I. L. *Izv. Akad. Nauk SSSR Ser. Khim.* **1972**, 1223.
149. Laurent, E.; Marquet, B.; Tardivel, R.; Thiebault, H. *Tetrahedron Lett.* **1987**, *28*, 2359.
150. Laurent, E.; Marquet, B.; Tardivel, R. *J. Fluorine Chem.* **1990**, *49*, 115.
151. Laurent, E.; Marquet, B.; Tardivel, R. *Tetrahedron* **1989**, *45*, 4431.
152. Koch, V. R.; Miller, L. L.; Clark, D. B.; Fleishmann, M.; Joslin, T.; Pletcher, D. *J. Electroanal. Chem.* **1973**, *43*, 318.
153. Meurs, J. H. H.; Eilenberg, W. *Tetrahedron* **1991**, *47*, 705.
154. Andres, D. F.; Laurent, E. G.; Marquet, B. S.; Benotmane, H.; Bensadat, A. *Tetrahedron* **1995**, *51*, 2605.
155. Laurent, E. G.; Tardivel, R.; Benotmane, H.; Bensadat, A. *Bull. Soc. Chim. Fr.* **1990**, *127*, 468.
156. Hara, S.; Chen, S. Q.; Toshio, T; Fukuhara, T.; Yoneda, N. *Tetrahedron Lett.* **1996**, *37*, 8511.
157. Laurent, R.; Tardivel, R.; Thiebault, H. *Tetrahedron Lett.* **1983**, *24*, 903.
158. Ventalon, F. M.; Faure, R.; Laurent, E. G.; Marquet, B. S. *Tetrahedron Asymmetry* **1994**, *5*, 1909.
159. Chen, S. Q.; Hatakeyama, T.; Fukuhara, T.; Hara, S.; Yoneda, N. *Electrochim. Acta* **1997**, *42*, 1951.
160. Fuchigami, T.; Konno, A.; Nakagawa, K.; Shimoji, M. *J. Org. Chem.* **1994**, *59*, 5933.
161. Fuchigami, T.; Shimojo, M.; Konno, A. *J. Org. Chem.* **1995**, *60*, 3459.
162. Brigaud, T.; Laurent, E. *Tetrahedron Lett.* **1990**, *31*, 2287.
163. Baroux, P.; Tardrel, R.; Simonet, J. *J. Electrochem. Soc.* **1997**, *44*, 841.
164. Narizuka, S.; Koshiyama, H.; Konno, A.; Fuchigami, T. *J. Fluorine Chem.* **1995**, *73*, 121.
165. Fuchigami, T.; Hayashi, T.; Konno, A. *Tetrahedron Lett.* **1992**, *33*, 3161.
166. Fuchigami, T.; Hayashi, T. Unpublished results.
167. Fuchigami, T.; Fujita, T.; Konno, A. *Tetrahedron Lett.* **1994**, *35*, 4157.

168. Aliev, I. Y.; Rozhkov, I. N.; Knunyants, I. L. *Tetrahedron Lett.* **1976**, *17*, 2469.
169. Fuchigami, T.; Miyazaki, M. *Electrochim. Acta* **1997**, *42*, 1979.
170. Nikitin, E. V.; Ignat'ev, Y. A.; Romakhim, A. S.; Parakin, O. V.; Romanov, G. V.; Kargin, Y. M.; Pudovik, A. N. *Zh. Obshch. Khim.* **1982**, *52*, 1207.
171. Nikitin, E. V.; Ignat'ev, Y. A.; Romakhim, A. S.; Parakin, O. V.; Kosachev, I. P.; Romanov, G. V.; Kargin, Y. M.; Pudovik, A. N. *Zh. Obshch. Khim.* **1982**, *52*, 2792.
172. Nikitin, E. V.; Ignat'ev, Y. A.; Parakin, O. V.; Kargin, Y. M.; Pudovik, A. N. *Zh. Obshch. Khim.* **1983**, *52*, 230.
173. Nikitin, E. V.; Kazakova, A. A.; Parakin, O. V.; Kargin, Y. M.; Pudovik, A. N. *Zh. Obshch. Khim.* **1982**, *52*, 2027.
174. Nikitin, E. V.; Parakin, O. V. *Izv. Akad. Nauk SSSR Ser. Khim.* **1986**, 1686.
175. Gambaretto, G. P.; Napoli, M.; Franccaro, C.; Conte, L. *J. Fluorine Chem.* **1982**, *19*, 427.
176. Ballinger, J. R.; Teare, F. W. *Electrochim. Acta* **1985**, *30*, 1075.
177. Makino, K.; Yoshioka, H. *J. Fluorine Chem.* **1988**, *30*, 1075.
178. Meurs, J. H. H.; Eilenberg, W. *Tetrahedron* **1991**, *47*, 705.
179. Sono, M.; Morita, N.; Shimizu, Y.; Tori, M. *Tetrahedron Lett.* **1994**, *35*, 9237.
180. Fuchigami, T. *Rev. Heteroatom. Chem.* **1994**, *10*, 155.
181. Narizuka, S.; Konno, A.; Matsuyama, H.; Fuchigami, T. *Denki Kagaku* **1993**, *61*, 868.
182. Narizuka, S.; Fuchigami, T. *Bioorg. Med. Chem. Lett.* **1993**, *5*, 1293.
183. Fuchigami, T.; Narizuka, S.; Konno, A.; Momota, K. *Electrochim. Acta* **1998**, *43*, 1985.
184. Higashiya, S.; Narizuka, S.; Konno, A.; Maeda, T.; Momota, K.; Fuchigami, T. *J. Org. Chem.* **1999**, *64*, 133.
185. Fuchigami, T.; Narizuka, S.; Konno, A. *J. Org. Chem.* **1992**, *57*, 3755.
186. Konno, A.; Naito, W.; Fuchigami, T. *Tetrahedron Lett.* **1992**, *33*, 7017.
187. Fuchigami, T.; Sakanashi, T. Unpublished results.
188. Narizuka, S.; Fuchigami, T. *J. Org. Chem.* **1993**, *58*, 4200.
189. Suda, K.; Hotoda, K.; Aoyagi, M.; Takanami, T. *J. Chem. Soc. PT. 1* **1995**, 1327.
190. Hou, Y.; Higashiya, S.; Fuchigami, T. *Synlett.* **1997**, 655.
191. Hou, Y.; Higashiya, S.; Fuchigami, T. *J. Org. Chem.* **1997**, *62*, 8773.
192. Andres, D. F.; Dietrich, U.; Laurent, E. G.; Marquet, B. S. *Tetrahedron* **1997**, *53*, 647.
193. Yoshida, J.; Ishichi, Y.; Tsoe, S. *J. Am. Chem. Soc.* **1992**, *114*, 7594.
194. Yoshida, J.; Sugawara, M.; Kise, N. *Tetrahedron Lett.* **1996**, *37*, 3157.
195. Erian, A. W.; Konno, A.; Fuchigami, T. *Tetrahedron Lett.* **1994**, *35*, 7245.
196. Erian, A. W.; Konno, A.; Fuchigami, T. *J. Org. Chem.* **1995**, *60*, 7654.
197. Hou, Y.; Higashiya, S.; Fuchigami, T. *J. Org. Chem.* **1997**, *62*, 9173.
198. Konno, A.; Fuchigami, T. *J. Org. Chem.* **1997**, *62*, 8579.
199. Yoshiyama, T.; Fuchigami, T. *Chem. Lett.* **1992**, 1995.
200. Fuchigami, T.; Sano, M. *J. Electroanal. Chem.* **1994**, *309*, 255.

201. Fuchigami, T.; Sano, M. *J. Electroanal. Chem.* **1995**, *414*, 81.
202. Fuchigami, T.; Fujita, T. *J. Org. Chem.* **1994**, *59*, 7190.
203. Fujita, T.; Fuchigami, T. *Tetrahedron Lett.* **1996**, *37*, 4725.
204. Fuchigami, T.; Mitomo, K. Unpublished results.
205. Fuchigami, T.; Tetsu, M.; Sakanashi, T. Unpublished results.
206. Umemoto, T.; Tomizawa, G. *Bull. Chem. Soc. Jpn.* **1986**, *59*, 3625.
207. Umemoto, T.; Tomizawa, G. *J. Org. Chem.* **1995**, *60*, 6563.
208. McCarthy, J. R.; Peet, N. P.; LeTourneau, M. E.; Inbasekaran, M. *J. Am. Chem. Soc.* **1985**, *107*, 735.

PHOTOINDUCED ELECTRON TRANSFER REACTIONS OF ORGANOSILICON COMPOUNDS

Kazuhiko Mizuno, Toshiyuki Tamai,
Akira Sugimoto, and Hajime Maeda

Advances in Electron Transfer Chemistry
Volume 6, pages 131–165.

ISBN: 0-7623-0213-5

1. INTRODUCTION

Photoinduced electron transfer (PET) reactions between electron-donating molecules (D) and electron-accepting molecules (A) serve as useful methods for the generation of reactive ionic species under neutral and mild conditions.[1–14] PET reactions generate a variety of reactive species depending on the electron-donating or electron-accepting ability of D and A, and the reaction media such as solvents and additives. It is well known that exciplexes, $[A^{\delta-}D^{\delta+}]^*$, are formed by weak charge transfer interactions between electronically excited A (or D), A* (or D*), with ground states of D (or A) in solvents of low polarity. The appearance of exciplex fluorescence supports the existence of exciplex intermediates. In polar solvents, efficient electron transfer takes place from D to A* (or from D* to A) to generate contact radical ion pairs, solvent-separated radical ions, and free radical ions.

The free energy changes associated with formation of radical ion species via PET reactions can be conveniently predicted by use of the

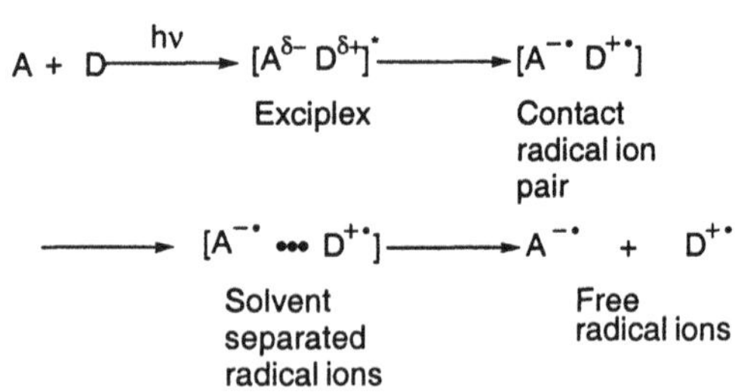

(1)

$$\Delta G = E_{1/2}^{ox}(D/D^{+}) - E_{1/2}^{red}(A^{-}/A) - E_{0\text{-}0} - e^{2}/\varepsilon r \qquad (2)$$

simple formulation shown in Eq. 2,[15] where $E^{ox}_{1/2}(D/D^{+})$, $E^{red}_{1/2}(A^{-}/A)$, E_{0-0}, and $e^2/\varepsilon r$ are the oxidation potential of D, the reduction potential of A, the 0–0 excitation energy of the excited molecule, and the Coulombic interaction energy between $D^{+\bullet}$ and $A^{-\bullet}$ in a solvent of dielectric constant ε of distance r apart, respectively. When ΔG is negative, the radical ions are produced in an exothermic process at near diffusion-controlled rates.

The photochemistry of organosilicon compounds has been extensively investigated not only from synthetic and mechanistic perspectives, but also with the intent of determining characteristic chemical and physical properties.[16–22] Very recently, Steinmetz reviewed the area of organosilane photochemistry in which he focused on the reactivities of mono-, di-, tri-, and polysilanes.[23] The most interesting and important point of that review is the differences in the reactivity that exist between organosilicon compounds and the corresponding all-carbon compounds.

The role of the silyl groups in organic chemistry can be simply classified by its use for (1) activation of functional groups, (2) protection of functional groups, and (3) providing steric bulk.[24–26] In the area of organic photochemistry, these characteristic aspects can be utilized to generate reactive species. Allylic and benzylic silanes are electron-donating molecules activated by silyl groups. This is explained by consideration of σ–π interactions. Silyl groups of silyl enol ethers, ketene silyl acetals, and cyclopropanone silyl acetals are protected forms of the carbonyl groups of aldehydes, ketones, and esters. Alcohols, phenols, and carboxylic acids can also be protected by silyl groups. In general, the oxidation potentials or ionization potentials of organosilicon compounds are much lower than those of the corresponding all-carbon compounds. Therefore, the organosilicon compounds can be utilized as electron donors in excited and ground state electron transfer processes.

This review concentrates on PET reactions of organosilicon compounds in which they serve as electron donors. Although greatest attention will be given to the reactions described, mechanistic aspects of these photoreactions will be described briefly.

2. PHOTOCHEMICAL ACTIVATION OF ORGANOSILICON COMPOUNDS: FORMATION AND REACTIVITIES OF EXCIPLEXES, RADICAL IONS, AND NEUTRAL RADICALS

Two major pathways operate in the photoreactions of organosilicon compounds, $XSiR_3$, with electron-accepting molecules, A. One is the formation of excited state complexes or exciplexes and this occurs in less polar solvents. The exciplex often gives rise to formation of cycloadducts without loss of the silyl group or to decay back to the ground state with or without exciplex fluorescence. The second pathway is the generation of radical ion species via single-electron transfer from the organosilicon compounds to A and this predominates in polar solvents. The radical ion species include contact radical ion pairs, solvent-separated radical ions, and free radical ions. The radical cations of the organosilicon compounds normally react to generate neutral radical and cation fragments via cleavage of C–Si bond or Si–Si bonds. In the case of monosilanes, radical cation fragmentation produces neutral carbon radicals and silyl cation components. On the other hand, both silyl radicals and silyl cation components are generated from the radical cations of di-, tri-, and oligosilanes. The cleavage of the organosilicon radical cations is promoted by nucleophiles such as acetonitrile, water, and alcohols. A preferred second step in these processes leads to formation of a new C–C bond via the addition of the radical $X^{\bullet}$ to the radical anion of the electron acceptor, $A^{-\bullet}$, to give the anionic adduct which then yields a neutral product by reactions such as protonation or elimination of anionic species. The other possible reaction pathways are the cross coupling of $X^{\bullet}$ with a radical generated by one-electron reduction of a cationic species of A such as an iminium salt, the homo-coupling of $X^{\bullet}$, and the reaction of $X^{\bullet}$ with molecular dioxygen.

Two examples of the PET reactions of allylic silanes with electron-accepting molecules are described here. Mizuno and his co-workers reported the photosubstitution of 1,4-dicyanobenzene (DCB) by allylic silanes as shown in Scheme 2.[27] The first step is one-electron transfer from the allylic silanes to the excited singlet state of 1,4-dicyanobenzene to generate the radical cation of allylic silanes and the radical anion of 1,4-dicyanobenzene. The second step is the cleavage

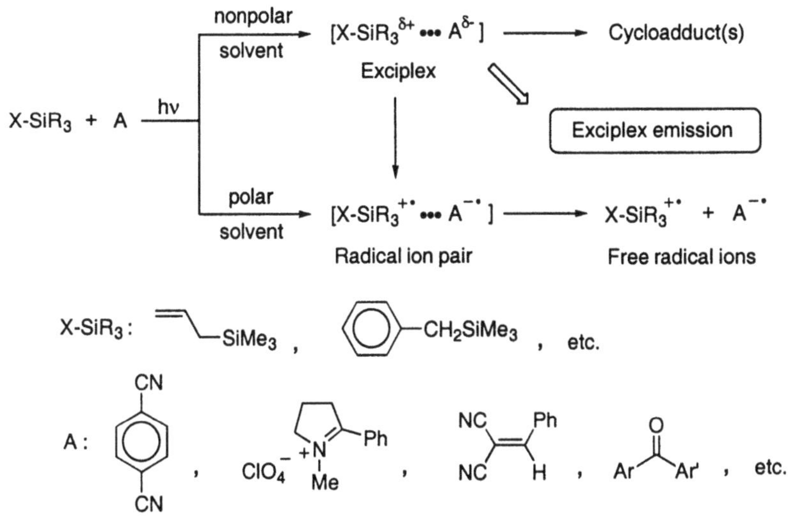

Scheme 1.

of the radical cations of allylic silanes to give allyl radical and a silyl cation component. The third step is addition of the allyl radical to the radical anion of 1,4-dicyanobenzene to give the anionic intermediate which is then followed by decyanation to produce 1-allyl-4-cyanobenzene.

It is notable that C–Si bond cleavage in the radical cation intermediate is assisted by attack of a nucleophile on a silyl group. This conclusion is supported by the observation that the quantum yield for the formation of 1-allyl-4-cyanobenzene decreases with the bulkiness of silyl group in the order $SiMe_3 > SiEt_3 > SiMe_2{}^tBu > Si^iPr_3 > SiPh_3$ (Table 1).[27] Very recently, Dinnocenzo and his co-workers also found that the rate for photocleavage of the C–Si bond of benzylsilane radical cations depends not only on the bulkiness of the silyl group but also on the bulkiness of the alcohol nucleophiles.[28] In addition, Mizuno showed that addition of aromatic solvents such as benzene and toluene to serve as π-acceptors enhances the quantum yields for the formation of the substitution product (Table 2).[27]

Also operating in PET reaction pathways is the rapid back electron transfer from $A^{-\bullet}$ to $D^{+\bullet}$.[9] This process which leads to reduced quantum efficiencies can be partly avoided by use of redox photosensitization

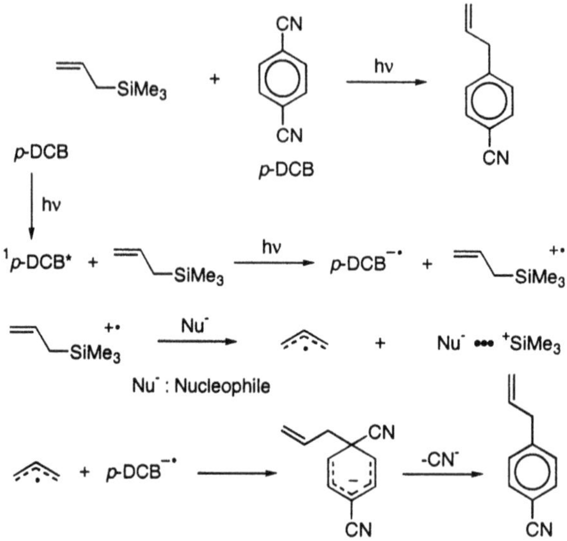

Scheme 2.

or cosensitization techniques. For example, the efficiency of PET reactions can be enhanced by the addition of aromatic hydrocarbons, ArH.[29] The role of ArH is through π-complexation of D with $ArH^{+\bullet}$, $[D\cdots ArH]^{+\bullet}$, as shown in Scheme 3. The primary process in the ArH-sensitized photoallylation of *p*-DCB is electron transfer reaction from the excited singlet ArH, $^1ArH^*$, to DCB to afford $ArH^{+\bullet}$ and

Table 1. Effect of Substituents Bonded to Silyl Groups on the Photoallylation of *p*-DCB with $CH_2{=}CHCH_2SiR_3$

Run	*R₃*	$E^{ox}_{p/2}$ *(V)*[a]	Φ[b]
1	Me_3	1.58	0.223 (1.00)
2	$Me_2\,{}^tBu$	2.06	0.055 (0.25)
3	Et_3	1.77	0.180 (0.82)
4	$^iPr_3$	1.90	0.040 (0.18)
5	Ph_3	1.63	0.037 (0.17)

Notes: [a]Oxidation potentials versus Ag/Ag^+ in CH_3CN.
[b]Quantum yields for the formation of 1-allyl-4-cyanobenzene. Relative quantum yields are shown in parentheses.

Table 2. Effect of Additives on the Photoallylation of *p*-DCB with Allyltrimethylsilane

Additive	I_p *(eV)*[a]	Φ_{rel}[b]
None	—	1.0
Benzene	9.24	4.9
Toluene	8.82	6.3
m-Xylene	8.58	9.2
o-Xylene	8.56	10.4
p-Xylene	8.44	8.7
Mesitylene	8.40	11.9
Anisole	8.21	10.0

Notes: [a]Ionization potentials.
[b]Relative quantum yields for formation of allyltrimethylsilane. CH_3CN: additive = 3:1.

$DCB^{-\bullet}$. Secondary electron transfer then occurs from the allylsilane D to $ArH^{+\bullet}$ or the π-complexation of $ArH^{+\bullet}$ with D, $[ArH \cdots D]^{+\bullet}$. The radical cation species $D^{+\bullet}$ or $[ArH \cdots D]^{+\bullet}$ reacts to produce the allyl radical, which reacts with *p*-$DCB^{-\bullet}$ to give the substitution product shown in Scheme 2. Aromatic hydrocarbons such as phenanthrene, naphthalene, triphenylene, and *p*-terphenyl (Table 3) are effective in the redox photosensitization in the photosubstitution of DCB by allyltrimethylsilane.[27]

A second example is found in the photoaddition of allylic or benzylic silanes to iminium salts reported by Mariano and his coworkers.[30,31] PET reactions of allylic or benzylic silanes with iminium salts generate two kinds of neutral radicals, allyl or benzyl radicals and α-amino radicals. The C–C bond-forming reaction between two different radicals occurs by cross coupling (Scheme 4).

The product distribution in the photoreactions of organosilicon compounds with electron-deficient compounds often depends on the reaction media.[32–35] In nonpolar solvents, the photocycloaddition of allylsilanes to electron-deficient aromatic compounds occurs to give cyclobutanes via exciplex intermediates. In contrast, photoallylation of the aromatic compounds or the carbonyl group takes place via free radical ions in polar solvents. Some examples are shown in Schemes 5 and 6.

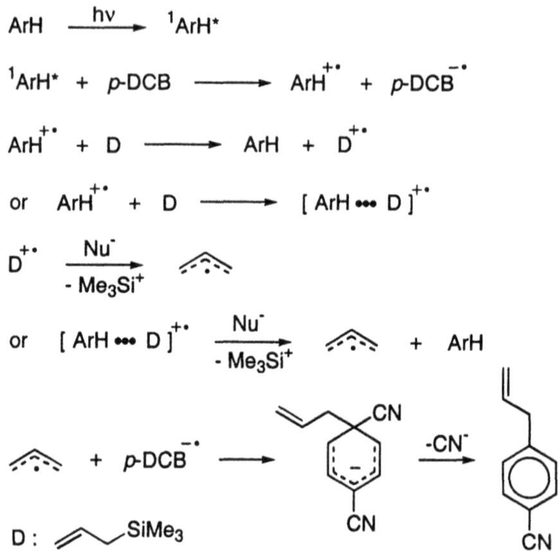

Scheme 3.

Table 3. Effect of Sensitizers on Photoallylation of *p*-DCB with Allyltrimethylsilane

Sensitizer	$E^{ox}_{p/2}$ *(V)*[a]	$\Delta E_{0,0}$ *(kJ mol⁻¹)*[b]	ΔG *(kJ)*[c]	Φ_{rel}[d]
None	—			1.0
p-Terphenyl	1.32			3.9
Triphenylene	1.29	348.9	–44.5	7.0
Naphthalene	1.22	384.9	–87.3	6.4
Phenanthrene	1.17	346.9	–54.0	3.6
Pyrene	0.78	322.2	–57.0	0.66
Anthracene	0.63	319.2	–78.5	0.93

Notes: [a]Oxidation potentials versus Ag/Ag^+ in acetonitrile. $E^{ox}_{p/2}$ of allyltrimethylsilane is 1.58 V.

[b]Singlet excitation energy.

[c]Free energy changes estimated by Rehm–Weller equation for a single electron transfer process from the excited singlet aromatic hydrocarbons, $^1ArH^*$, to the ground state *p*-DCB ($E^{red}_{p/2}$ = –1.92 V).

[d]Relative quantum yields for formation of 1-allyl-4-cyanobenzene.

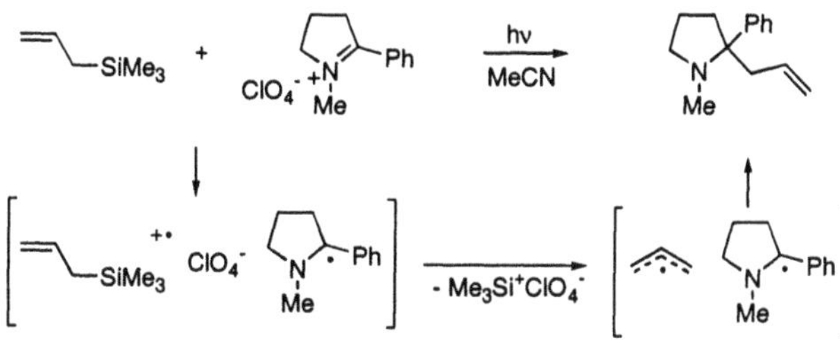

Scheme 4.

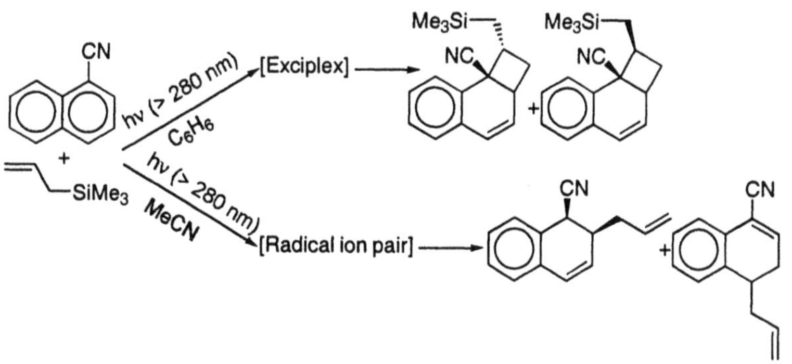

Scheme 5.

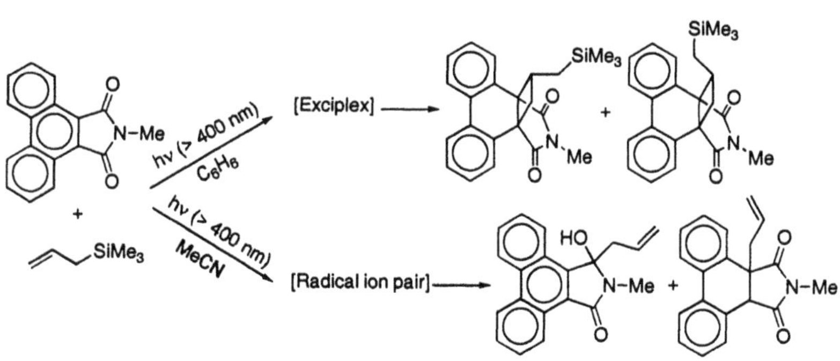

Scheme 6.

3. PHOTOREACTIONS OF ORGANOSILICON COMPOUNDS VIA PHOTOINDUCED ELECTRON TRANSFER

3.1. Photocycloaddition via Exciplexes

Photochemical 2π+2π-cycloaddition between electron donors and electron acceptors to give cyclobutanes is a popular pathway in organic photochemistry. Organosilicon compounds having unsaturation, such as allylsilanes and silyl enol ethers, can be utilized as electron-rich alkenes, because their oxidation potentials are about the same magnitude as those of alkyl vinyl ethers. Therefore, these substrates generally react with electron-deficient alkenes or aromatic compounds via exciplexes, giving cyclobutanes. The ring opening of these cycloadducts assisted by Lewis or Brønsted acids would be expected to give rise to allylated or Michael addition products as shown in Scheme 7.

Ochiai and colleagues previously reported the 2π+2π-photocycloaddition of the triplet excited state of 1,4-naphthoquinone with allyltrimethylsilane.[36] Similar photocycloadditions of 2-trimethylsilylcyclopentenone and 5-trimethylsilyluracil with alkenes have been discussed by Swenton and colleagues.[37,38] The photocycloadditions of allyltrimethylsilane to electron-deficient aromatic compounds via exciplexes have also been reported by other groups.[32–35]

Mizuno, Pac, and co-workers reported photo-Michael reactions via the 2π+2π photocycloaddition–hydrolysis sequences of silyl enol ethers with electron-deficient alkenes such as acrylonitrile, methyl acrylate, and 1-cyanonaphthalene.[39,40] Similarly, regioselective 2π+2π-photocycloaddition of 1-trimethylsiloxynaphthalene with

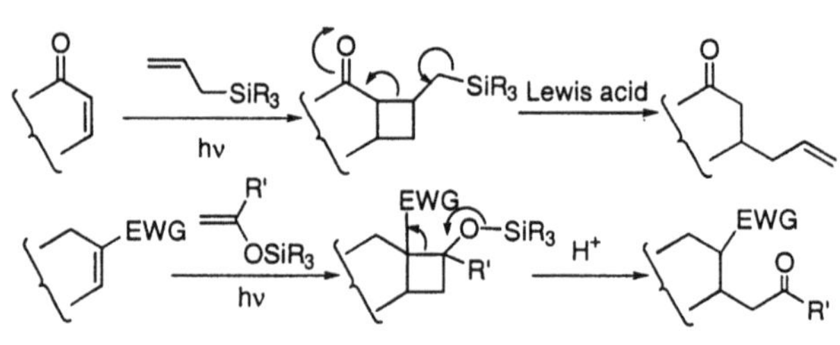

Scheme 7.

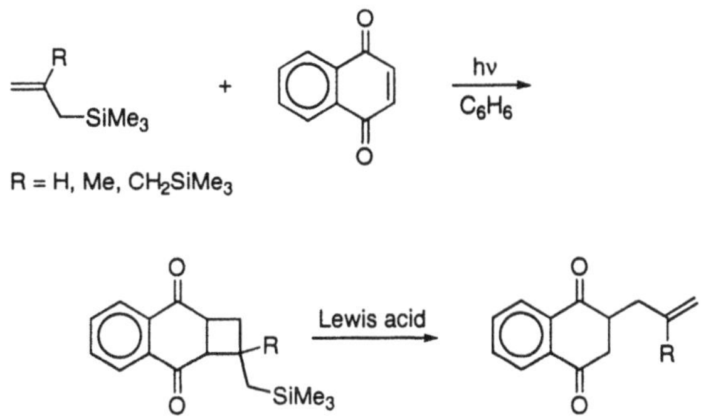

(3)

acrylonitrile occurs to give cyclobutanes which are readily hydrolyzed to form Michael-type adducts. Very recently, it has been reported that photo-Michael adducts are directly obtained by the photoreactions of silyl enol ethers with 1,1-dicyano-2-phenylethenes.[41] Trimethylsilylmethylnaphthalenes react with acrylonitrile to give *endo*-selective $2\pi+2\pi$-photocycloadducts. In these photoreactions, triplet or singlet exciplexes between organosilicon compounds and electron-accepting molecules are postulated to serve as the key reactive intermediates.

Allylic silanes,[42] silyl enol ethers,[43–47] and ketene silyl acetals[48,49] serve as electron-rich alkenes in photoreactions with a variety of carbonyl compounds, such as benzophenone, to give oxetanes in a regio- and stereoselective manner.[50] The ring-opening reactions of

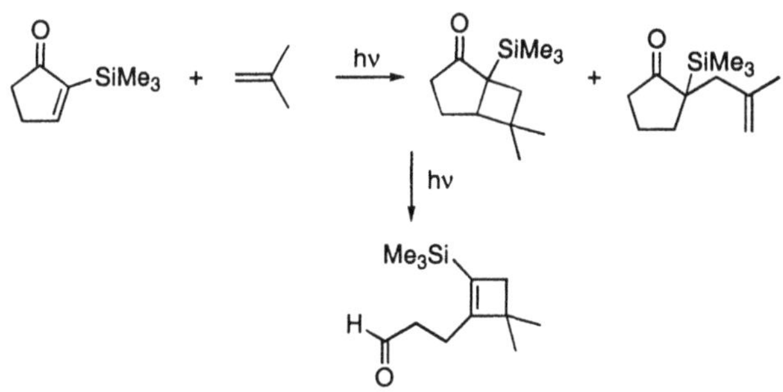

(4)

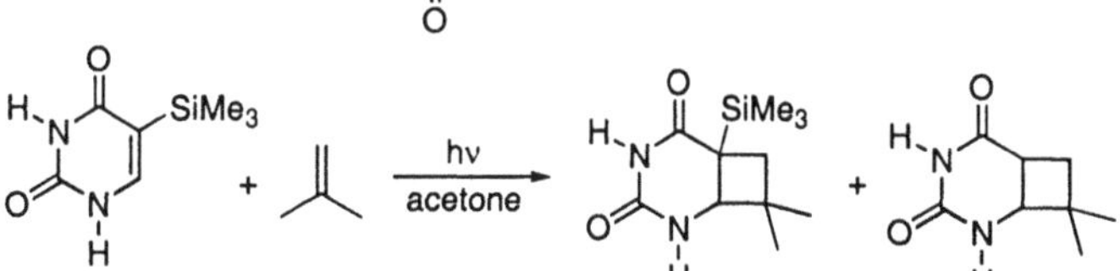

(5)

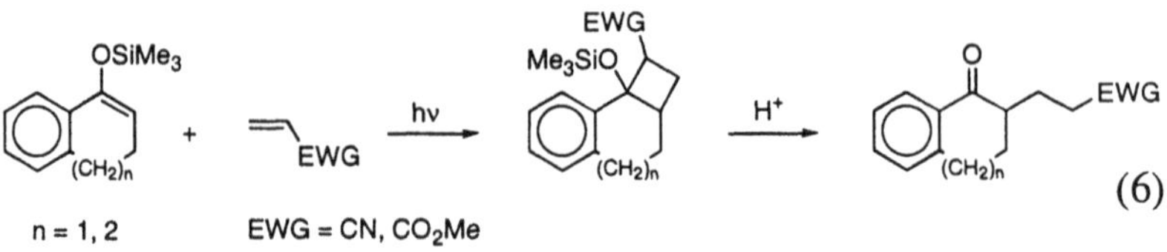

(6)

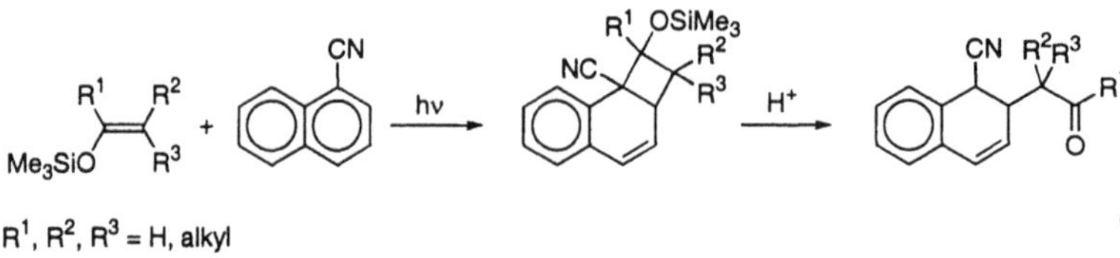

(7)

(8)

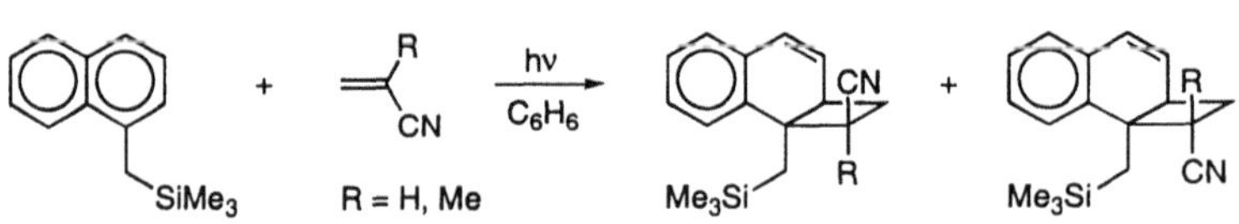

(9)

these oxetanes have been developed to give polyfunctionalized acyclic molecules.

The intramolecular photocyclodimerization of anthracenes and aryl-substituted alkenes having silyl groups via excimer intermediates has been reported by several groups.[51,52] In these photoreactions, the silyl groups either activate alkenyl or aryl groups or act as a spacer. In some cases, two alkenyl groups are temporarily connected by a silyl

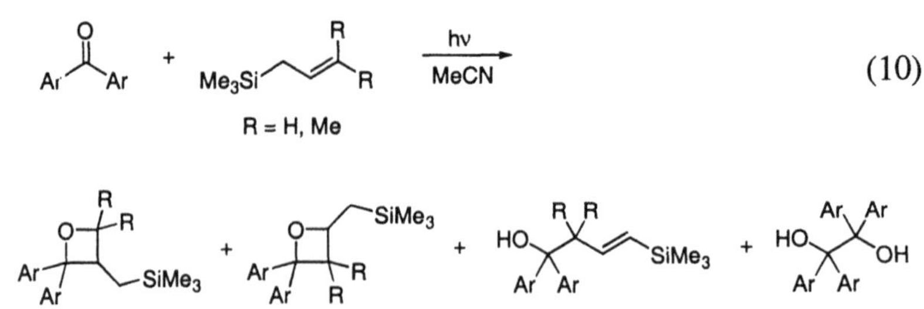

(10)

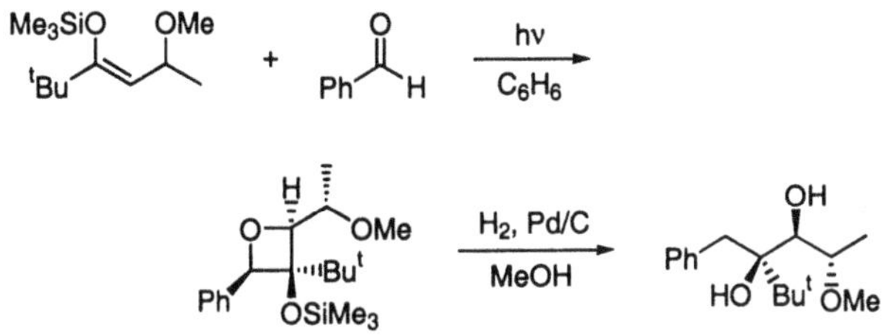
Me3SiO
OMe
tBu
+
O
Ph
H
hν
C6H6
H
OMe
O
Bu^t
Ph
OSiMe3
H2, Pd/C
MeOH
OH
Ph
HO
Bu^t
OMe

(11)

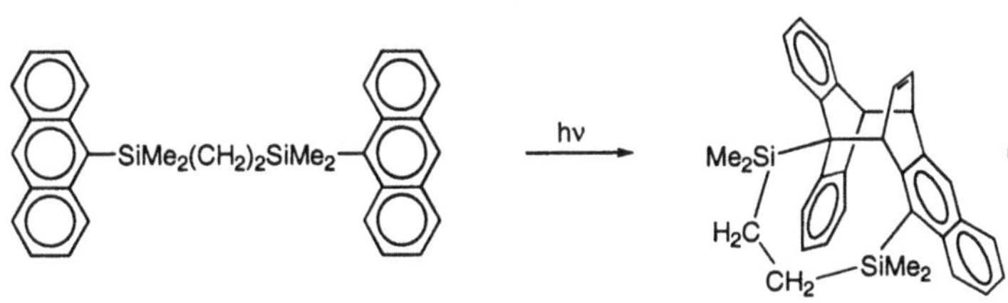
SiMe2(CH2)2SiMe2
hν
Me2Si
H2C
CH2
SiMe2

(12)

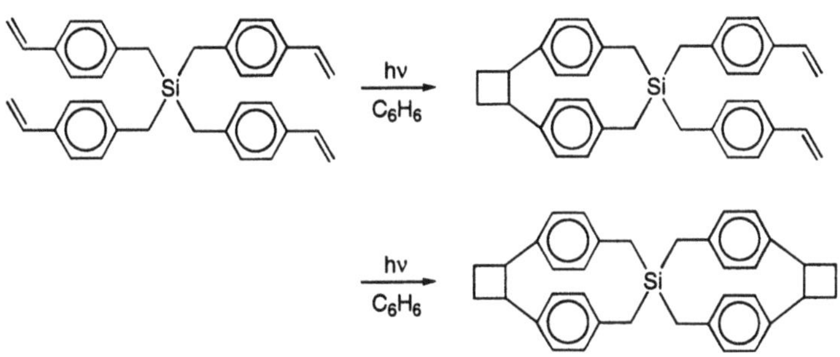
Si
hν
C6H6
Si
hν
C6H6
Si

(13)

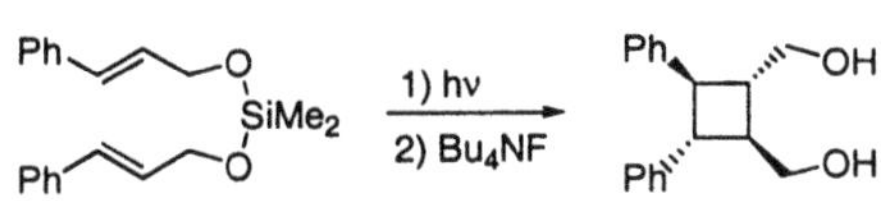
Ph
O
SiMe2
Ph
O
1) hν
2) Bu4NF
Ph
OH
Ph
OH

(14)

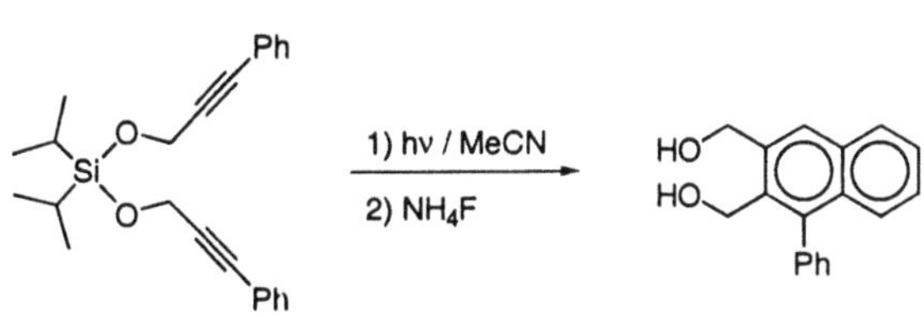
Ph
Si
O
O
Ph
1) hν / MeCN
2) NH4F
HO
HO
Ph

(15)

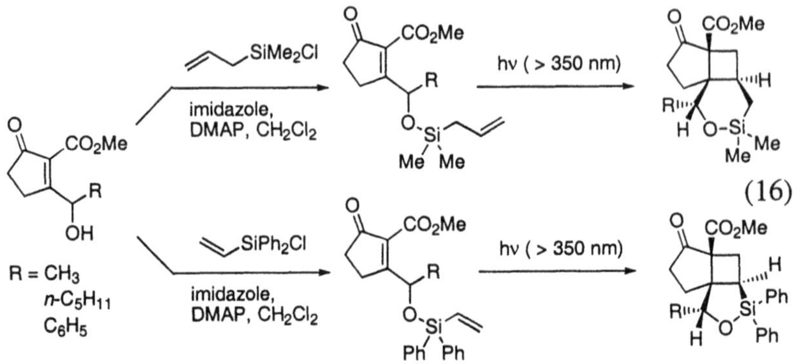

(16)

group and easily unconnected by hydrolysis after the intramolecular photocycloaddition.[53–55]

3.2. Photocycloaddition via Triplexes

Termolecular interactions in excited states are an interesting aspect of recent organic photochemical studies. In general, terminal alkenes and alkyl vinyl ethers do not easily dimerize to give cyclobutanes under direct or sensitized photoreaction conditions. However, 1,4-dicyanonaphthalene (DCN) can be used as an excellent sensitizer for the dimerization of alkyl vinyl ethers and trimethylsiloxyethene in benzene or toluene.[56] In this process, the triplex ^{1}[DCN···toluene···vinyl ether]* generated from DCN, toluene, and vinyl ether has been postulated as a key reactive intermediate based on the observation that the

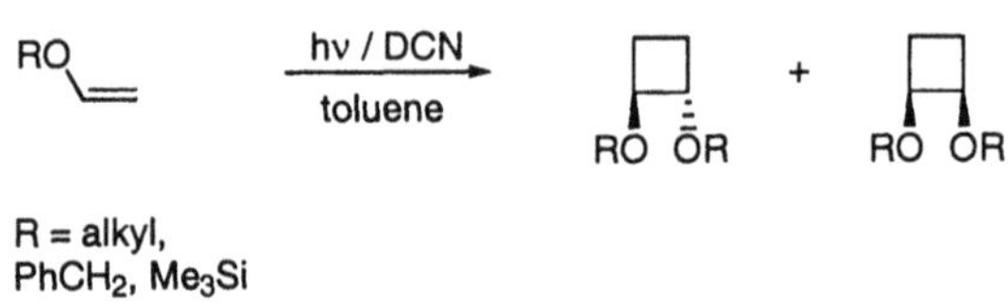

(17)

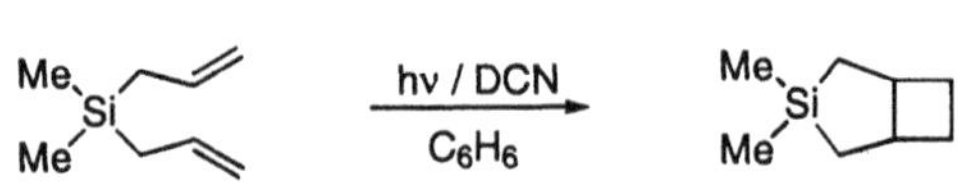

(18)

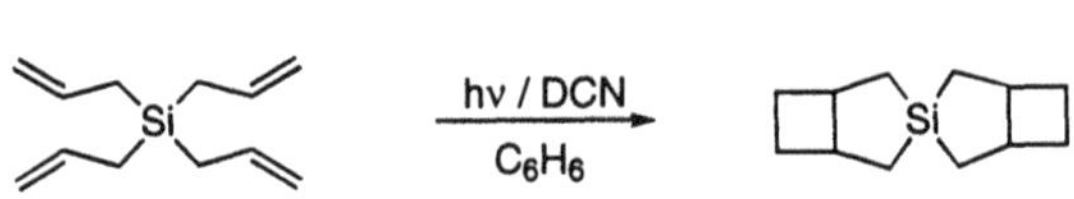

(19)

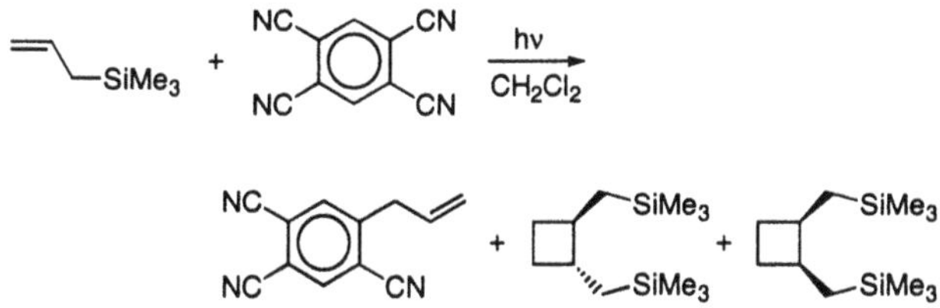

(20)

lifetime of DCN in toluene is much longer than in cyclohexane and that the exciplex emission between DCN and toluene is efficiently quenched by vinyl ethers. This technique has been applied to DCN-sensitized intramolecular photocycloaddition of diallylsilanes and tetraallylsilane, giving 3-silabicyclo[3.2.0]heptanes and related spiro compounds, respectively.[57] Albini and co-workers reported the intermolecular photocyclodimerization of allyltrimethylsilane when 1,2,4,5-tetracyanobenzene is used as an electron acceptor.[34]

3.3. Photosubstitution of Aromatic Compounds

Photosubstitution of polycyano-aromatic compounds has been achieved by use of organosilicon compounds such as allylic silanes and arylmethylsilanes, and in some cases using alkyl silanes. The primary process in these reactions is the generation of neutral radicals $R^{\bullet}$ via radical cations of organosilicon compounds as described above. Addition of $R^{\bullet}$ to the radical anions of polycyano-aromatic com-

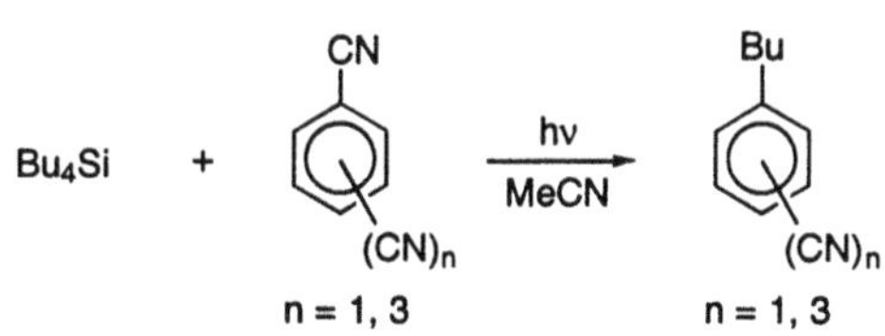

(21)

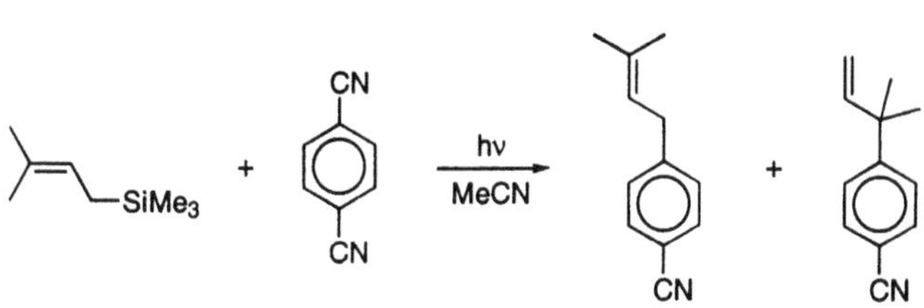

(22)

2 : 3

pounds produces anionic intermediates, which are converted into substitution and addition products.

The photosubstitution of 1,4-dicyanobenzene by use of allyltrimethylsilane in acetonitrile giving 1-allyl-4-cyanobenzene is a typical example as shown in Scheme 2.[27] Similar photoreactions of 1,2- and 1,4-dicyanobenzenes and 1,2,4,5-tetracyanobenzene using tetraalkylsilanes,[58,59] benzylsilanes,[60] and disilanes[61] occur to give substitution products regioselectively. The photoreaction of 1,4-dicyanobenzene with 3-methyl-2-butenyltrimethylsilane gives both α- and γ-allylated products in a 2:3 ratio.[27,60] This ratio is not dependent on the solvent used or the presence of additives. These results clearly support a mechanism involving formation of allyl and benzyl radicals as intermediates.

Electron-donating silyl ethers such as ketene silyl acetals and cyclopropanone silyl acetals also can be used as latent carbon-functional groups.[62] The photoreactions of 1,2,4,5-tetracyanobenzene with dimethyl ketene silyl acetal and 2,2-dimethylcyclopropanone silyl methyl acetal afford substituted products in good yields, respectively.

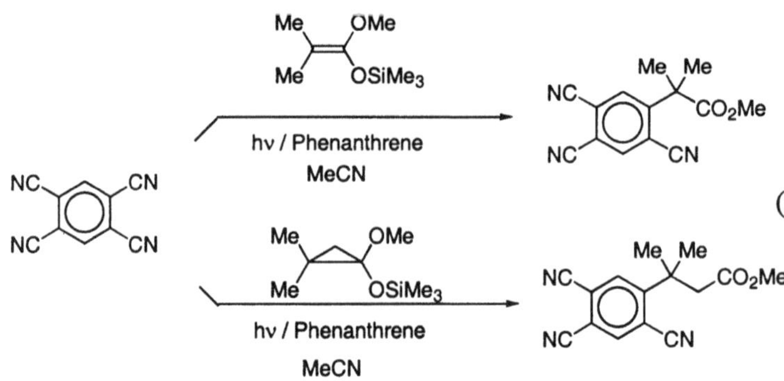

(23)

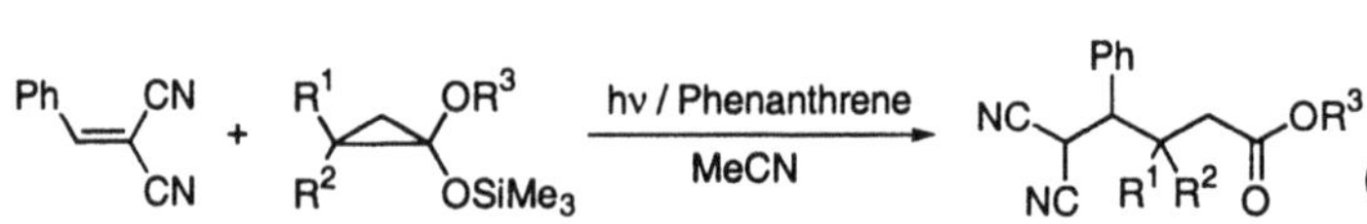

(24)

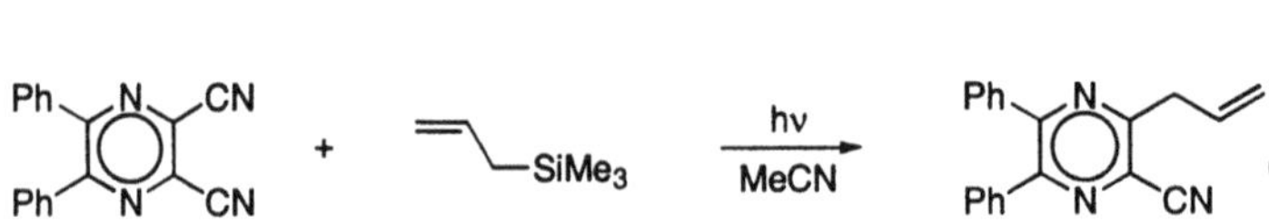

(25)

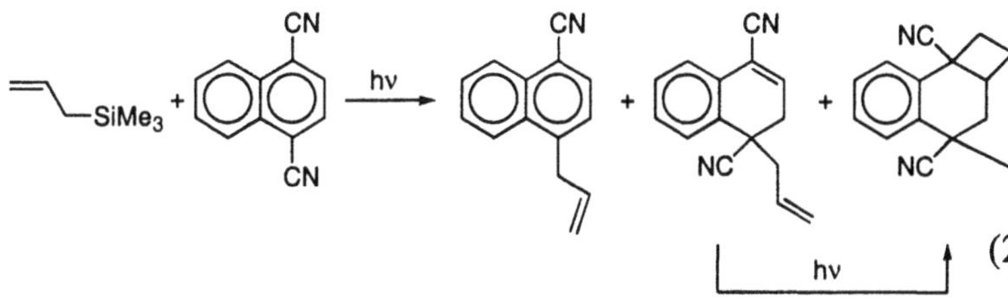

(26)

in MeCN / $NaOCH_3$	98%		
in MeCN / EtOH / pyrene	trace	78%	
in MeCN / EtOH / phenanthrene	19%		72%

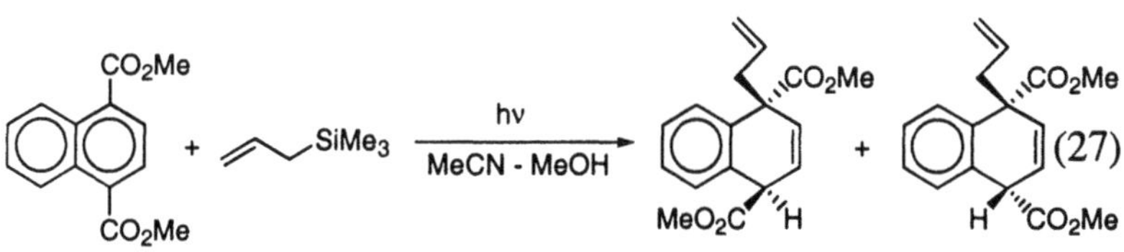

(27)

One of the cyano groups of 3,4-dicyanopyrazine is also substituted by allylic and benzylic silanes and ketene silyl acetals to give allylated, benzylated, and alkylated products in high yields.[63,64]

The photoreactions of electron-deficient polyaromatic compounds with allylic silanes afford reductive photoallylation products efficiently. Mono- and di-cyanonaphthalenes regioselectively reacted with allyltrimethylsilane to give reductive allylation products.[35,65,66] Similar photoreactions of naphthalene dicarboxylates with allyl-

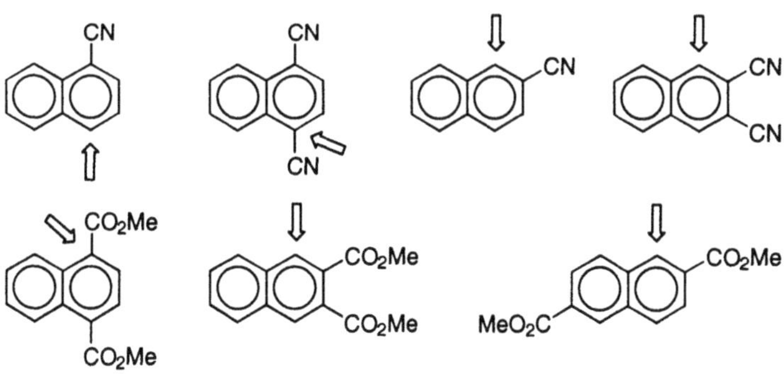

Scheme 8.

trimethylsilane afforded the reductive allylation products in a regioselective manner.[67] The regioselectivity for the initial allyl radical addition step can be argued based on spin densities in the radical anions of these electron-deficient naphthalene derivatives. The intramolecular 2π+2π-photocycloaddition of the reductive allylation products affords the benzotricyclo[4.2.1.0^{3,8}]nonene derivatives in high yields (Eq. 26).[35]

3.4. Photoaddition to Electron-Deficient Unsaturated Compounds and Arenes

In 1982, Ohga and Mariano first reported the photoallylation of iminium salts.[30] The photoreaction of iminium salts with allyltrimethylsilane in acetonitrile affords allylated compounds via the addition of an allyl radical to an α-aminomethyl radical.[68] Allyl radicals are produced via the cleavage of C–Si bonds of allylsilane radical cations and the α-aminomethyl radicals are generated by one-electron reduction of iminium salts as shown in Scheme 4.

The photoreaction of 1,1-dicyano-2-phenylethene with allylsilane in the presence of phenanthrene in acetonitrile affords 5,5-dicyano-4-phenyl-1-pentene in good yield.[69] Phenanthrene acts as a redox sensitizer in this process and allylation occurs at the position β to the cyano group. On the other hand, a similar photoallylation reaction of cyclo-

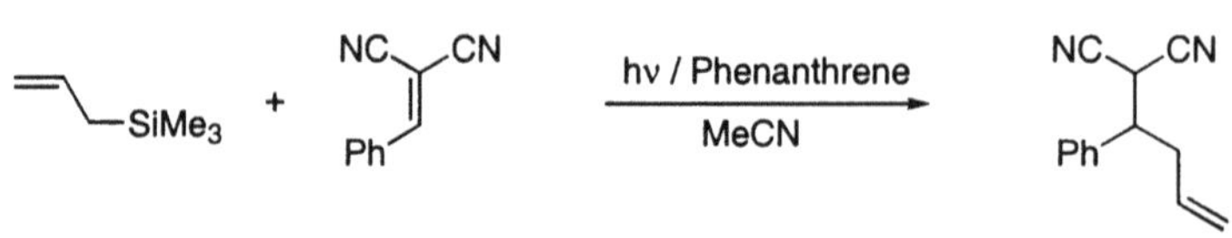

(28)

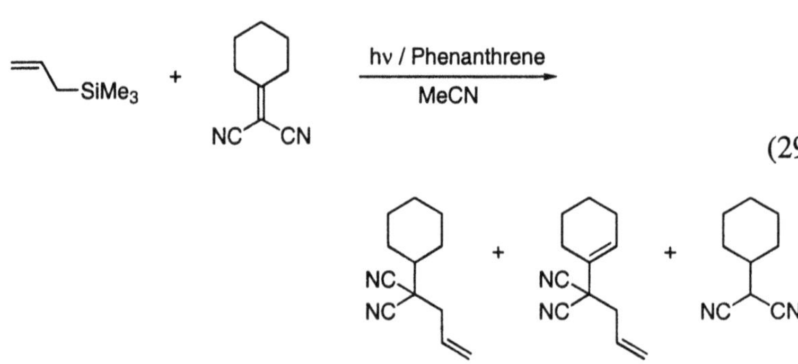

(29)

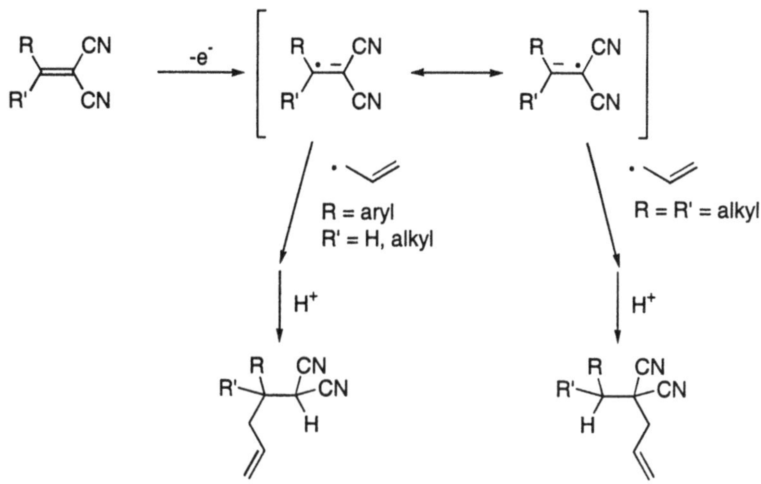

Scheme 9.

hexylidenepropanedinitrile occurs at the position α to the cyano group. The difference between the aryl- and alkyl-substituted dinitriles regarding the regioselectivity can be attributed to the difference in the electronic structures of their radical anions. In the case of the aryl-substituted dinitriles, the radical is localized on the carbon to which the aryl group is attached and the radical center is stabilized by conjugation with the aryl group. In contrast, the radical of alkyl-substituted dinitriles would be localized on the carbon to which the nitrile groups are attached. The unusual electronic distribution in the latter radical anion comes from capto-dative stabilization of the radical.[70,71] The radical center is stabilized by the interaction with electron-withdrawing cyano groups and the electron-donating carbanionic moiety.

Reductive arylmethylation of DCN and 9,10-dicyanoanthracene (DCA) by use of arylmethylsilanes via PET pathways has been reported by Mizuno, Albini, and Dinnocenzo, independently.[72–79]

Fukuzumi and his co-workers reported the photoreaction of 2-naphthaldehyde with allylic silanes in the presence of $Mg(ClO_4)_2$ which gives allylated products.[80] In the absence of $Mg(ClO_4)_2$, this photoreaction does not occur. The interaction of Mg^{2+} with 2-naphthaldehyde makes its singlet lifetime much longer. Similarly,

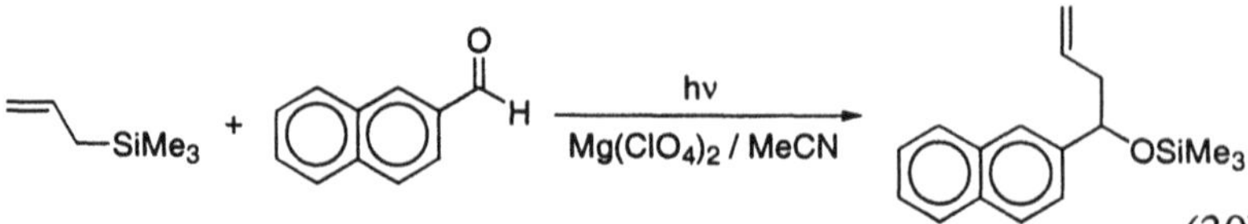

(30)

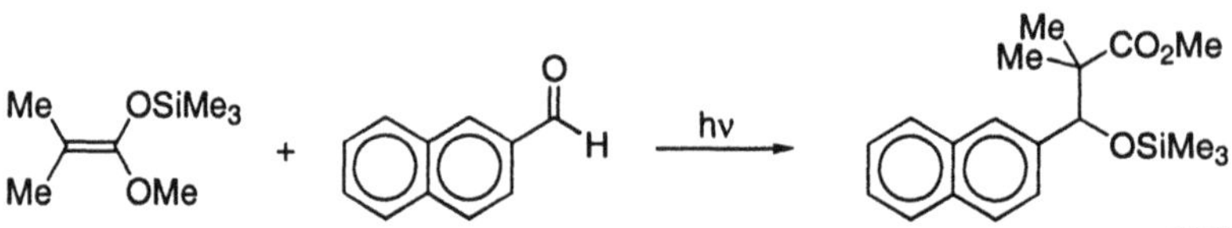

(31)

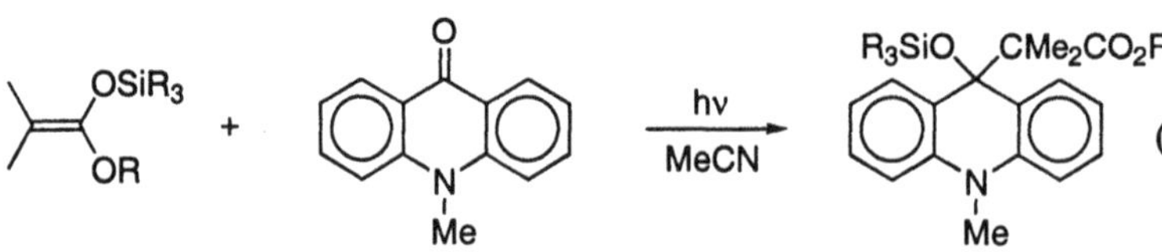

(32)

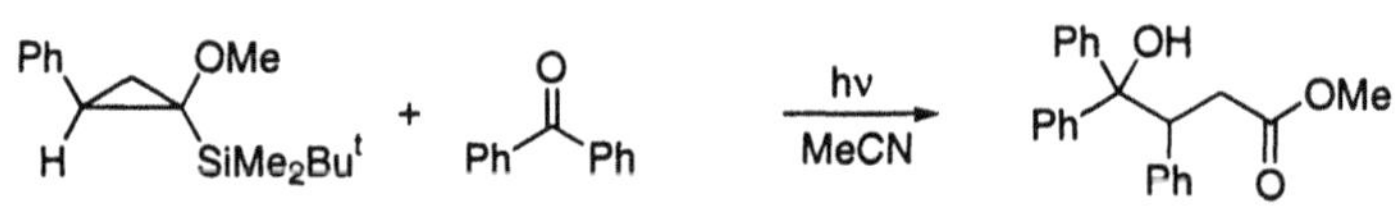

(33)

Mg^{2+} catalyzes photoalkylations of 2-naphthaldehyde using ketene silyl acetals which give alkylated adducts in good yields.

PET reactions of aryl ketones with ketene silyl acetals and cyclopropanone silyl acetals in acetonitrile afford β-hydroxy- and γ-hydroxy-carbonyl compounds in good yields.[49,81–83]

3.5. Photocyclization

Intramolecular photocyclization is a useful method for the preparation of cyclic and heterocyclic compounds. Mariano and his coworkers have developed intramolecular PET reactions of iminium salts having silyl groups and used this process to prepare selected natural products.[31] Intramolecular electron transfer from the organosilicon group to the iminium salt generates the diradical cation. The intramolecular radical coupling of the 1,*n*-diradical intermediate formed via desilylation gives *N*-heterocyclic products.

Mariano also reported several examples of inter- and intramolecular C–C bond-forming reactions using α-aminomethylsilyl compounds

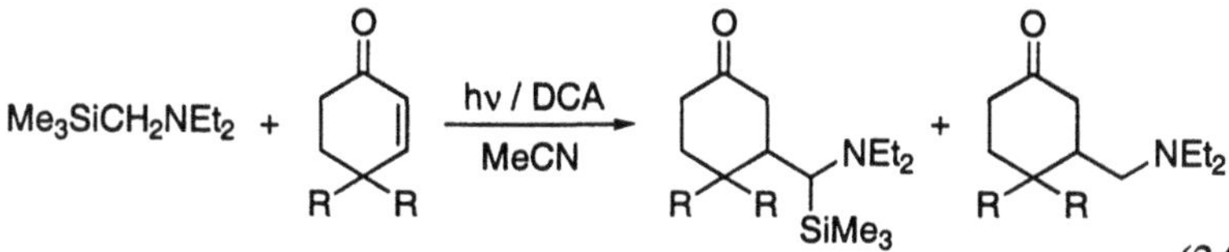

(34)

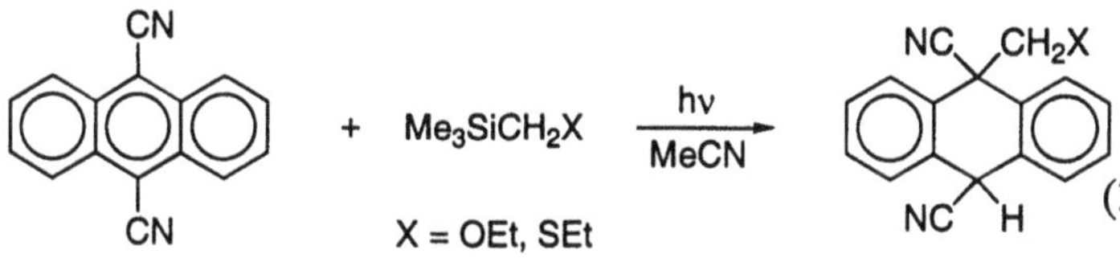

(35)

to α,β-unsaturated enones, DCA, and related compounds.[84–89] Yoon and colleagues have independently reported the intramolecular electron transfer-induced photocyclizations of phthalimide bearing silyl groups to give heterocyclic compounds.[90]

In these photoreactions, silyl groups are activated by an α-heteroatom and smoothly removed in reactions of the radical cation intermediates. DCA-sensitized photocyclization of cyclic silyl enol ethers has

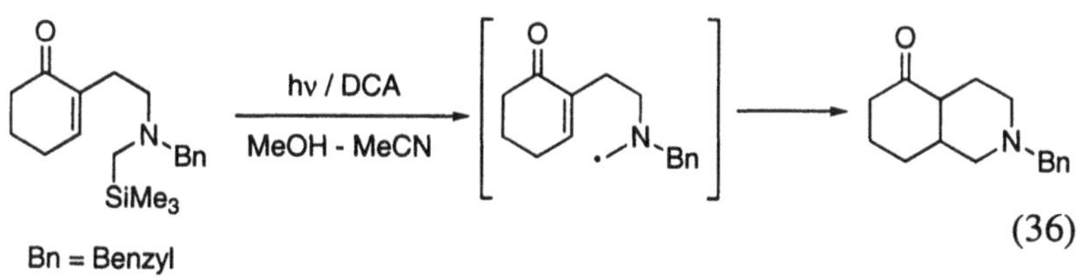

(36)

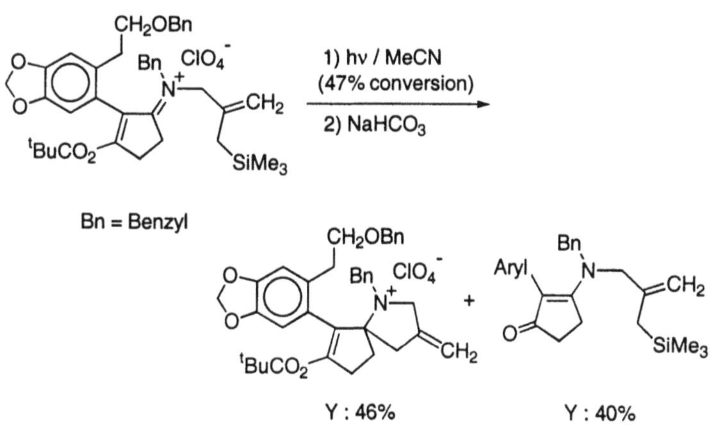

(37)

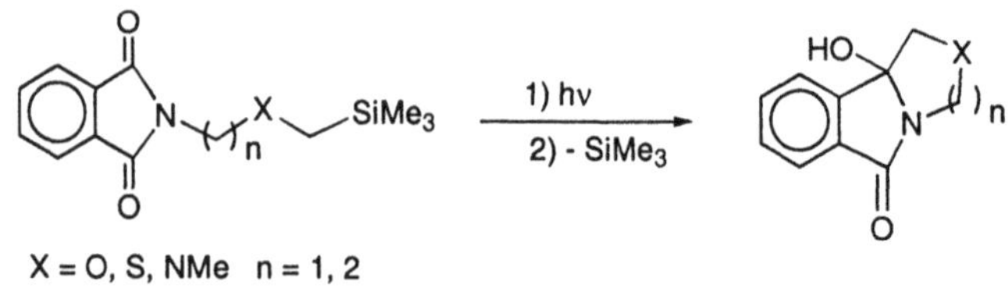

(38)

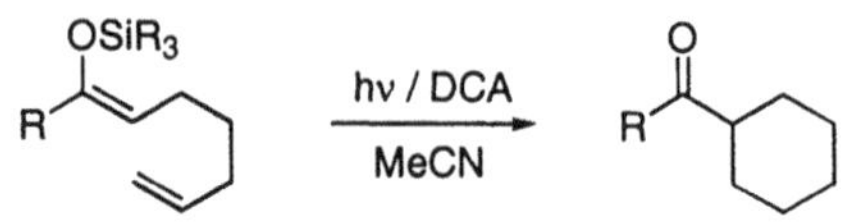

(39)

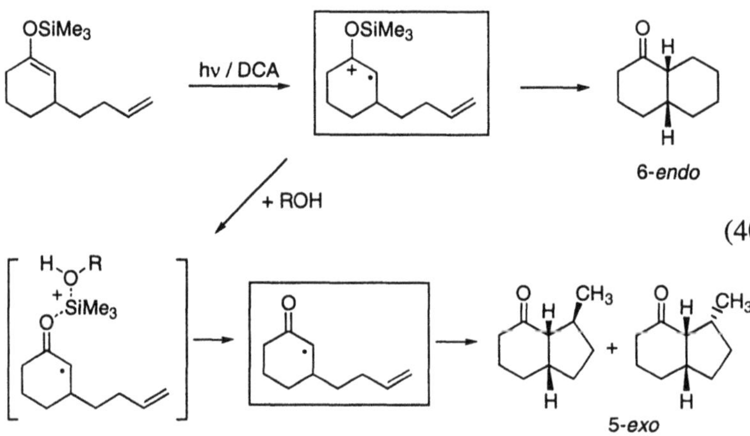

(40)

DCA : 9,10-dicyanoanthracene

been reported by Mattay and co-workers.[91,92] The key step in these processes is desilylation of the silyl enol ether radical cations to generate α-keto radicals.

3.6. Photosilylation Using Di-, Tri-, and Oligosilanes

In photoreactions of di-, tri-, and oligosilanes in the presence of acceptors (A), initiation is by electron transfer from the silanes to A and Si–Si bond cleavage occurs to give silyl radicals and silyl cation components. The silyl radicals display reactivity similar to that of allyl and arylmethyl radicals with the radical anions of the electron acceptors. The result is the introduction of silyl groups into the electron-deficient alkenes and arenes. When nucleophiles such as alcohols are

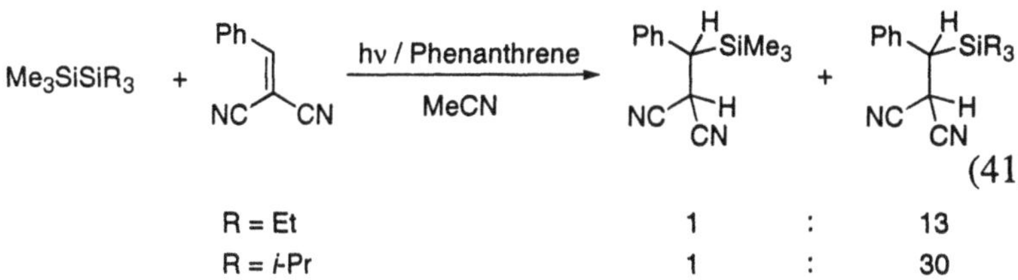

(41)

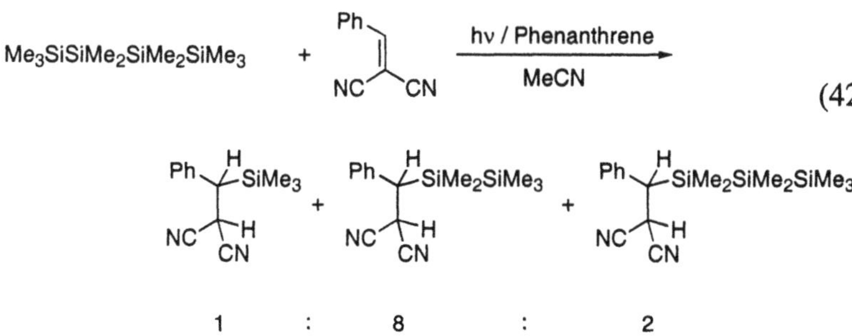

(42)

present in the reaction system, the silyl cations are trapped by the nucleophiles.[94,99]

Mizuno and his co-workers reported the photosilylation of electron-deficient alkenes by use of hexamethyldisilane and unsymmetrical disilanes.[93,94] The photoreaction of 1,1-dicyano-2-phenylethene and disilanes in acetonitrile in the presence of phenanthrene affords β-silylated dicyanoethanes in good yields. A key intermediate in this process is the silyl radical, generated by solvent (nucleophile)-assisted cleavage of the radical cation of disilanes. Silyl radical attack at the radical

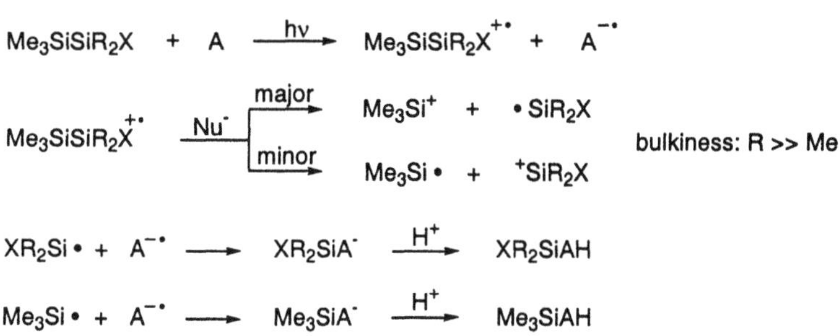

Scheme 10.

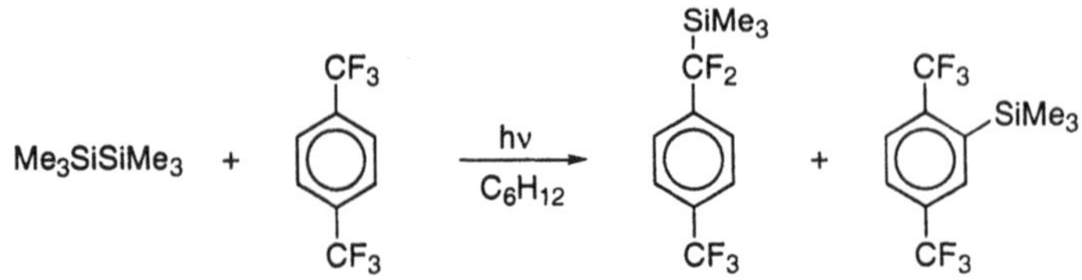

(43)

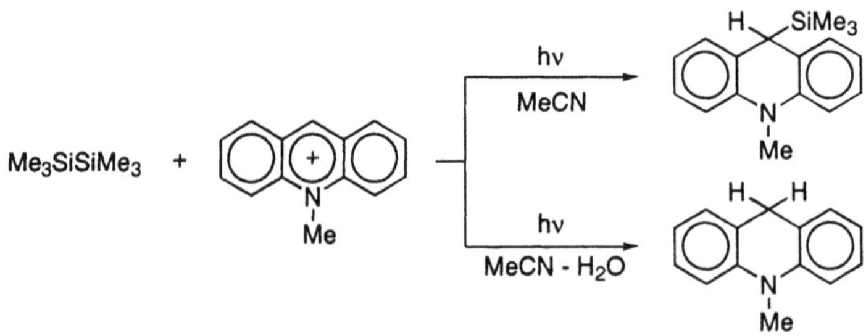

(44)

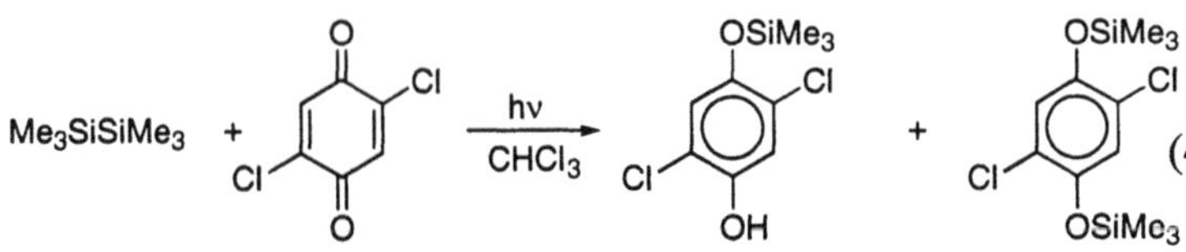

(45)

site of dicyanoethene radical anion then occurs. The radical cations of unsymmetrical disilanes undergo nucleophilic attack at the less-hindered Si atom to give the bulkier silyl radicals. Therefore, the bulkier silyl group is introduced to the dicyanoethenes.

Similar photosilylation reactions occur when one uses tri- and oligosilanes. Nakadaira and his co-workers reported the photosilylation of polycyano-aromatics and CF_3-substituted benzenes.[61,95] Fukuzumi's group also has shown that acridinium ions are also silylated in photoreactions with hexamethyldisilane.[96,97] The photoreaction of hexamethyldisilane with quinones occurs to give silylated hydroquinones.[98]

Silyl radicals participate as the silylation agents in the above reactions. However, silyl cations have not yet been characterized as intermediates in these processes. Nakadaira and his co-workers reported intramolecular trapping of silyl cation by use of a OH group.[99] In the photoreaction of cyclic oligosilane, ring-opening or insertion reactions at Si–Si bonds occur via Si–Si bond cleavage. The photore-

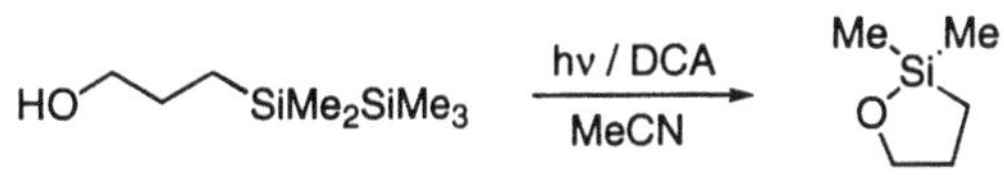

(46)

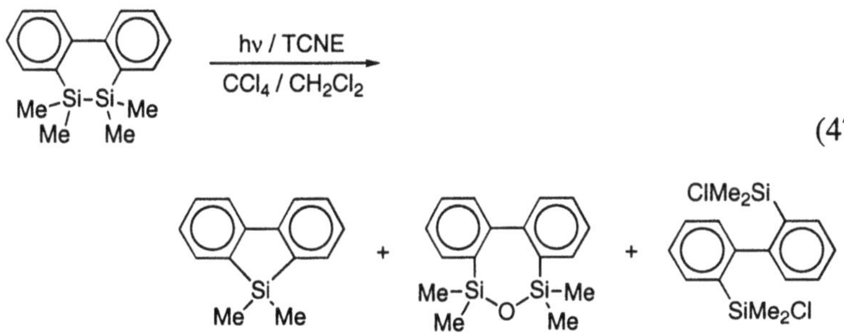

(47)

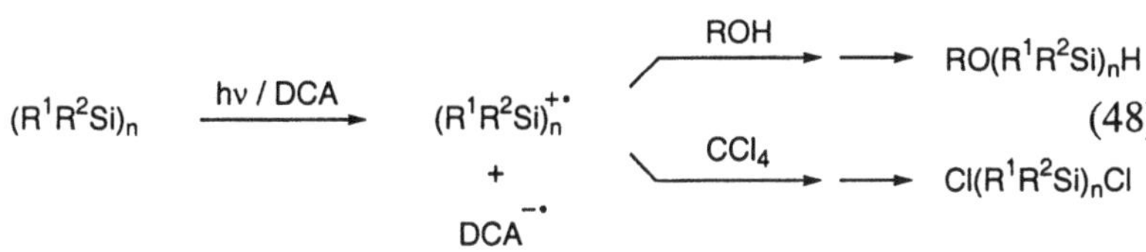

(48)

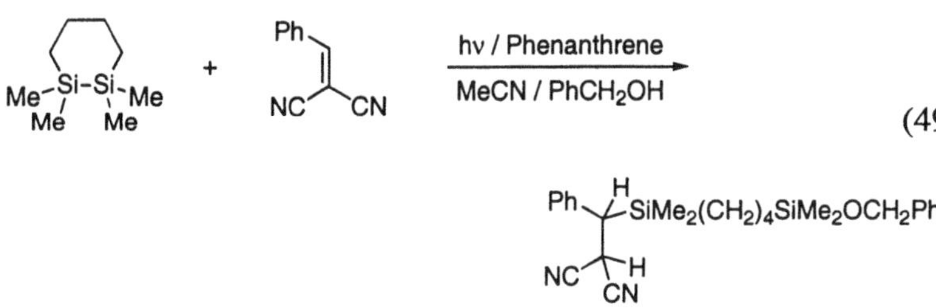

(49)

action of dibenzotetramethyldisilacyclohexadiene with TCNE in a mixed solvent system of dichloromethane and carbon tetrachloride (2:1) gives desilylated, oxygen-inserted, and Si-chlorinated products.[100] DCA-sensitized photoreaction of cyclic oligosilane in the presence of ethanol and carbon tetrachloride gives α-ethoxy-ω-hydropolysilane and α,ω-dichloropolysilane, respectively.[101,102] The phenanthrene-sensitized photoreaction of 1,2-disilacyclohexane with 1,1-dicyano-2-phenylethene in the presence of benzyl alcohol gives silylated products in which a benzyloxy group is incorporated at the

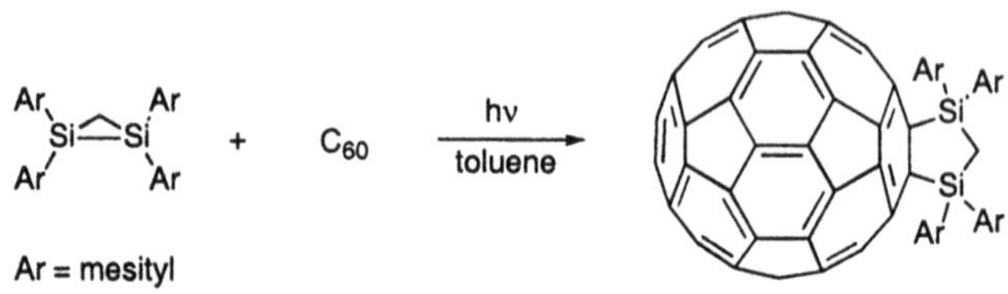

(50)

terminal silyl group.[94] In these photoreactions, both the silyl radical and silyl cation components are trapped by radicals and nucleophiles, respectively.

The photocycloadditions of small-ring disilanes to C-60 have been shown to give [2+3] and [2+4] adducts via exciplex intermediates.[103,104] C-60 also reacts with oligosilanes and ketene silyl acetals to afford photoaddition products.[105,106]

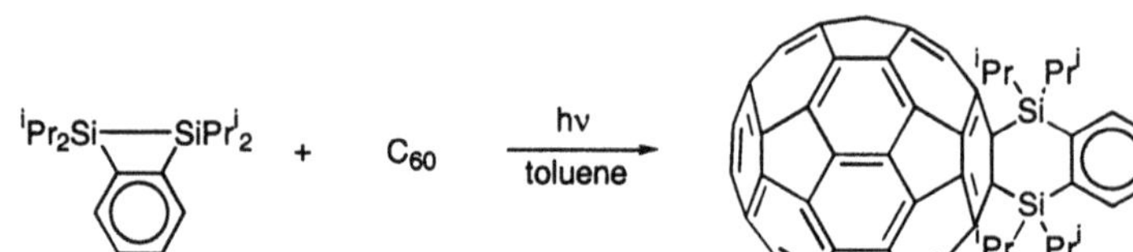

(51)

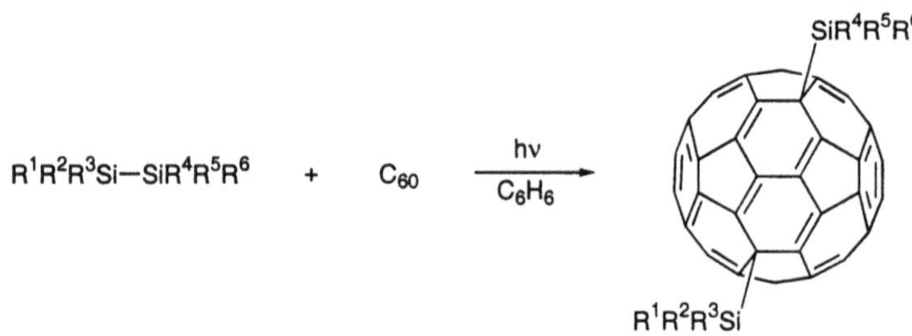

(52)

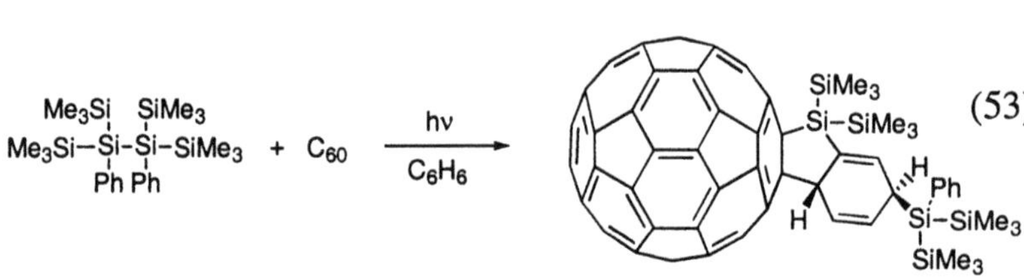

(53)

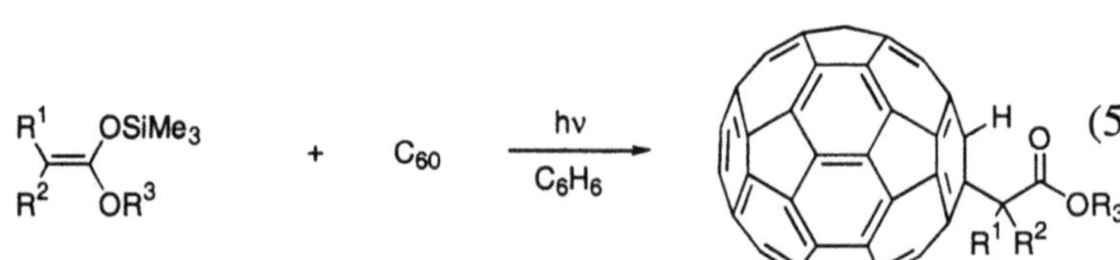

(54)

3.7. Photooxygenation of Organosilicon Compounds

It is well known that organic radicals are easily oxidized by molecular dioxygen. In PET reactions of organosilicon compounds, carbon or silyl radicals are generated by the cleavage of C–Si or Si–Si bonds in radical cations. Mizuno and his co-workers reported the DCA-sensitized photooxygenation of arylmethylsilanes to give aromatic carboxylic acids and aldehydes accompanied by some 1,2-diaryl- ethanes.[72,73] In these photoreactions, products of aromatic ring oxidation are not observed. TiO_2-sensitized photoreaction of arylmethysilanes in the presence of molecular dioxygen also gives oxygenated products.[107] Similar photoreactions of arylmethylsilanes in the presence of silver sulfate in place of molecular dioxygen affords only 1,2-diarylethanes. The photooxidation of arylmethylsilanes in the presence of $Cu(BF_4)_2$ generates arylmethyl cations by two-electron oxidation routes, which are trapped by alcohols and acetonitrile.[108]

Photooxygenation of disilanes and polysilanes is known to give disiloxanes and related compounds, but the mechanisms of these processes are not clear. Mizuno and his co-workers have reported the DCA-sensitized photooxygenation of aryldisilanes via PET to give siloxanes.[109] In this process, the radical cation of arylpentamethyldisilane is attacked by a nucleophile (e.g., acetonitrile or water) to give a silyl radical and a silyl cation, which originate from aryldisilane radical cation. The silyl radical reacts with molecular dioxygen and silyl cation is trapped by water, both giving silanols. Siloxanes as isolable products are obtained by the bimolecular condensation of silanols. The nucleophilic attack of water was confirmed by use of $H_2{}^{18}O$.

$$ArCH_2SiMe_3 + O_2 \xrightarrow{h\nu / DCA} ArCO_2H + ArCHO \quad (55)$$

$$ArCH_2SiMe_3 + ROH \xrightarrow[MeCN]{h\nu / Cu(BF_4)_2} ArCH_2OR \quad (56)$$

$$ArSiMe_2SiMe_3 + O_2 \xrightarrow[MeCN]{h\nu / DCA} ArSiMe_2OSiMe_2Ar \quad (57)$$

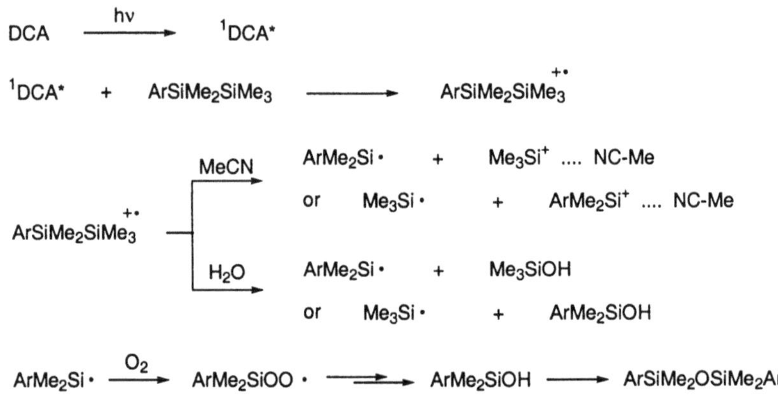

Scheme 11.

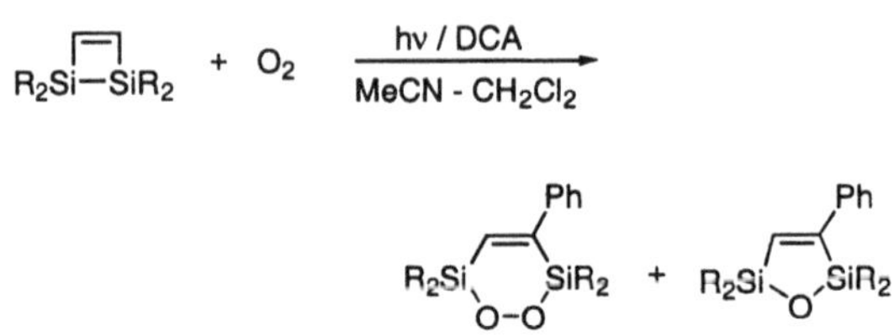

(58)

Ando and co-workers reported that electron transfer photooxygenation of three- and four membered cyclic disilanes gives cyclic peroxides.[110,111] DCA-sensitized photooxygenation of 1,2-disiletenes affords cyclic peroxides, 1,2,3,6-dioxadisilins, and 1,2,5-oxadisiloles. Similar photooxygenation of oxadisiliranes gives 1,2,4,3,5- trioxadisilolanes. For these photoreactions, Si–Si bond cleavage in the radical cations of the cyclic disilanes is postulated to be the key process.

3.8. Other Miscellaneous Reactions

PET reactions of benzylic silanes with polycyano-aromatic compounds have been applied to photoresist and electron-beam resist technologies.[112–114] The photoreaction of poly(4-trimethylsilylmethylstyrene) in benzene–acetonitrile with DCB affords insoluble polymer via photo-cross-linking, which contains 4-cyanophenylmethyl groups.[115,116] In the case of 1,2,4,5-tetracyanobenzene, soluble 2,4,5-tricyanophenylmethyl-substituted polystyrene is produced. But,

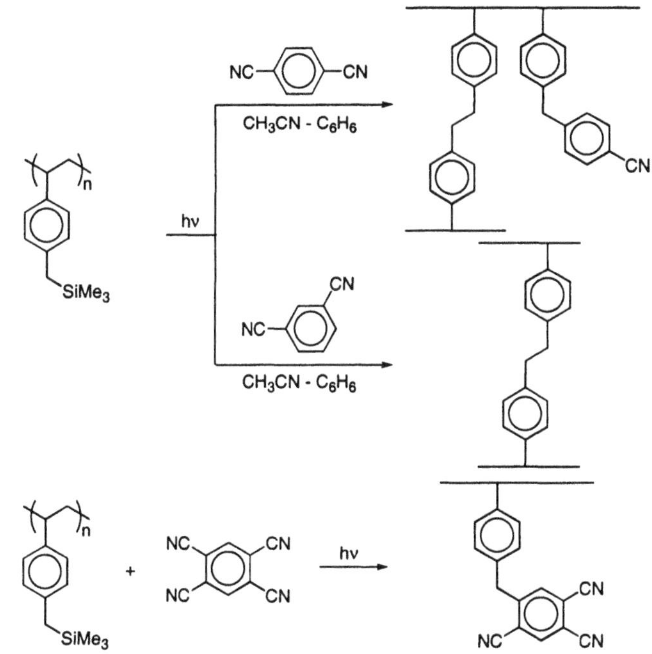

Scheme 12.

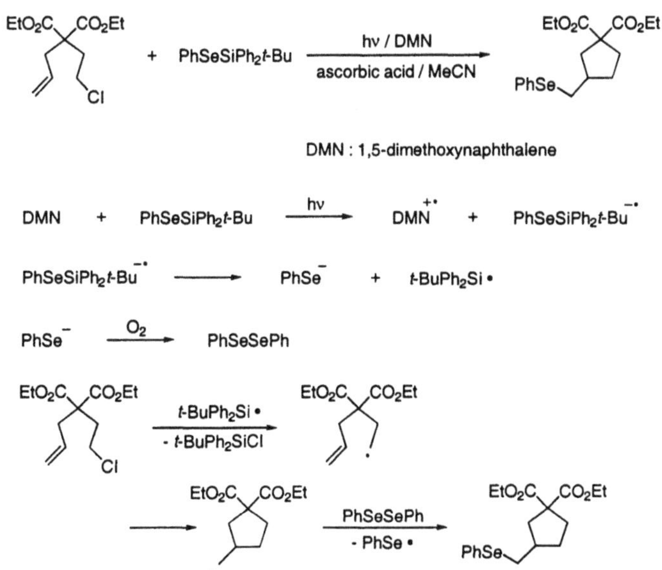

Scheme 13.

only cross-linked insoluble polymer is obtained in the case of 1,3-dicyanobenzene.

Pandey and colleagues reported the intramolecular photocyclization of δ-halo terminal alkenes using organosilicon compounds having Si–Se bond as shown in Scheme 13.[117] The 1,5-dimethoxynaphthalene-sensitized photoreaction generates $R_3Si^{\bullet}$ and $PhSe^-$ by the dissociation of the radical anion of PhSe–SiR_3. Key intermediates in this complex system are $R_3Si^{\bullet}$ and $PhSe^{\bullet}$. The former radical abstracts a halogen atom and the latter radical is trapped by a terminal radical giving a cyclopentane derivative.

4. CONCLUDING REMARKS

This review has covered many examples of PET reactions of organosilicon compounds in the presence of electron acceptors. In general, the SET-photochemical reactivities of organosilicon compounds can be classified into two categories. Monosilanes are common electron-donating organic compounds, which are protected or activated by monosilyl groups. The reactive species generated in the PET reaction pathways, depending on the reaction conditions, contain monosilyl groups. Both of the reactive species can be utilized for the construction of cyclic compounds or to induce C–C bond-forming reaction conditions. The advantage of organosilicon compounds is that the monosilyl group can be easily removed after or under the PET reactions. Organosilicon compounds having Si–Si bonds such as di- and trisilanes belong to the other category in terms of their PET reactions. At the present stage of development, disilanes and their related silanes can be used as silylating reagents via their radical cations.

The photooxygenation of organosilicon compounds is also an interesting subject from the synthetic and application viewpoints. The PET reactions of a polymer system having silyl groups and the organopolysilanes represent an attractive area in their applications.[118,119]

In PET reactions, a variety of reactive intermediates such as exciplexes and radical ions are postulated to explain the mechanism for the formation of photoproducts. There is a serious lack of direct evidence for the existence of these intermediates (except for Caldwell and Creed's previous work on exciplex systems).[120] The recent developments in the measurements of transient species such as radical ions using laser spectroscopy can provide a wealth of information about these transient intermediates. However, it is important to develop further understanding of the PET reactions from the mechanistic and synthetic viewpoints also. In this way, synthetic photochemists can design highly selective and efficient synthetic processes using characteristic properties of organosilicon compounds.

ACKNOWLEDGMENTS

K.M. gratefully acknowledges the contributions of his collaborators mentioned in the references and especially Emeritus Professors Yoshio Otsuji (Osaka Prefecture University) and Hiroshi Sakurai (Osaka University) and Dr. Chyongjin Pac (Kawamura Institute of Technology) for their continued encouragement in this field. Financial support has been provided by the Ministry of Education, Science, Sports, and Culture, Japan. We also thank Shin-Etsu Chemical Co., Ltd. for the gift of chlorotrimethylsilane and dichlorodimethylsilane.

REFERENCES

1. Chanon, M.; Hawley, M. D.; Fox, M. A. In *Photoinduced Electron Transfer*; Fox, M. A.; Chanon, M., Eds.; Elsevier: Amsterdam, 1988, Part D.
2. Mattay, J., Ed. *Photoinduced Electron Transfer I, II, III, IV, V*: (*Top. Curr. Chem.* **156, 158, 159, 163, 168**); *Electron Transfer I, II*: (*Top. Curr. Chem.* **169, 177**); Springer: Berlin, 1990–1996.
3. Mariano, P. S., Ed. *Advances in Electron Transfer Chemistry*; JAI Press: Greenwich, CT, 1991–1996; Vols. 1–5.
4. Mariano, P. S.; Stavinoha, J. L. In *Synthetic Organic Photochemistry*; Horspool, W. M., Ed.; Plenum: New York, 1984, p. 145.
5. Mattes, S. L.; Farid, S. *Org. Photochem.* **1983**, *6*, 233.
6. Davidson, R. S. In *Advances in Physical Organic Chemistry*, Vol. 19; Gold, V.; Bethel, D., Eds.; Academic: New York, 1983, p. 1.
7. Mattay, J. *Angew. Chem. Int. Ed. Engl.* **1987**, *26*, 825.
8. Kochi, J. K. *Angew. Chem. Int. Ed. Engl.* **1988**, *27*, 1227.
9. Mizuno, K.; Otsuji, Y. *Top. Curr. Chem.* **1994**, *169*, 301.
10. Kavarnos, G. J.; Turro, N. J. *Chem. Rev.* **1986**, *86*, 401.
11. Mattay, J. *Synthesis* **1989**, 233.
12. Raabe, G.; Michl, J. *Chem. Rev.* **1985**, *85*, 419.
13. West, R. *Angew. Chem. Int. Ed. Engl.* **1987**, *26*, 1201.
14. Mariano, P. S. *Org. Photochem.* **1987**, *9*, 1.

15. Rehm, D.; Weller, A. *Isr. J. Chem.* **1970**, *8*, 259.
16. Brook, A. G. In *The Chemistry of Organic Silicon Compounds*; Patai, S.; Rappoport, Z., Eds.; Wiley: New York, 1989, Part 2, Chapter 15.
17. Ishikawa, M.; Kumada, M. *Adv. Organomet. Chem.* **1981**, *19*, 51.
18. Raabe, G.; Michl, J. In *The Chemistry of Organic Silicon Compounds*; Patai, S.; Rappoport, Z., Eds.; Wiley: New York, 1989, Part 2, Chapter 17.
19. Brook, A. G.; Baines, K. M. *Adv. Organomet. Chem.* **1986**, *25*, 1.
20. Miller, R. D.; Michl, J. *Chem. Rev.* **1989**, *89*, 1359.
21. West, R. *Pure Appl. Chem.* **1984**, *56*, 163.
22. Sakurai, H. *J. Organomet. Chem.* **1980**, *200*, 216.
23. Steinmetz, M. G. *Chem. Rev.* **1995**, *95*, 1527.
24. Patai, S.; Rappoport, Z., Eds. *The Chemistry of Organic Silicon Compounds*, Parts 1 and 2; Wiley: New York, 1989.
25. Colvin, E. W., Ed. *Silicon in Organic Synthesis*; Butterworths: London, 1980.
26. Weber, W. P., Ed. *Silicon Reagents for Organic Synthesis*; Springer-Verlag: Berlin, 1983.
27. Nakanishi, K.; Mizuno, K.; Otsuji, Y. *Bull. Chem. Soc. Jpn.* **1993**, *66*, 2371.
28. Dockery, K. P.; Dinnocenzo, J. P.; Farid, S.; Goodman, J. L.; Gould, I. R.; Todd, W. P. *J. Am. Chem. Soc.* **1997**, *119*, 1876.
29. Majima, T.; Pac, C.; Nakasone, A.; Sakurai, H. *J. Am. Chem. Soc.* **1981**, *103*, 4499.
30. Ohga, K.; Mariano, P. S. *J. Am. Chem. Soc.* **1982**, *104*, 617.
31. Lan, A. J. Y.; Quillen, S. L.; Heuckeroth, R. O.; Mariano, P. S. *J. Am. Chem. Soc.* **1984**, *106*, 6439.
32. Kubo, Y.; Imaoka, T.; Shiragami, T.; Araki, T. *Chem. Lett.* **1986**, 1749.
33. Mizuno, K.; Yasueda, M.; Otsuji, Y. In *Symposium on Photochemistry*; Abstract IIA105, 1987, p. 35.
34. Mella, M.; Fasani, E.; Albini, A. *J. Org. Chem.* **1992**, *57*, 6210.
35. Nishiyama, T.; Mizuno, K.; Otsuji, Y.; Inoue, H. *Tetrahedron* **1995**, *51*, 6695.
36. Ochiai, M.; Arimoto, M.; Fujita, E. *J. Chem. Soc. Chem. Commun.* **1981**, 460.
37. Swenton, J. S.; Fritzen, E. L., Jr. *Tetrahedron Lett.* **1979**, 1951.
38. Shih, C.; Fritzen, E. L.; Swenton, J. S. *J. Org. Chem.* **1980**, *45*, 4462.
39. Mizuno, K.; Okamoto, H.; Pac, C.; Sakurai, H.; Murai, S.; Sonoda, N. *Chem. Lett.* **1975**, 237.
40. Pac, C.; Mizuno, K.; Okamoto, H.; Sakurai, H. *Synthesis* **1978**, 589.
41. Ramkumar, D.; Sankararaman, S. *J. Chem. Soc. Perkin Trans. 2* **1996**, 939.
42. Takuwa, A.; Fujii, N.; Tagawa, H.; Iwamoto, H. *Bull. Chem. Soc. Jpn.* **1989**, *62*, 336.
43. Bach, T.; Jödicke, K.; Kather, K.; Fröhlich, R. *J. Am. Chem. Soc.* **1997**, *119*, 2437.
44. Bach, T. *Liebigs Ann. Recueil* **1997**, 1627.
45. Bach, T.; Eilers, F.; Kather, K. *Liebigs Ann. Recueil* **1997**, 1529.
46. Bach, T. *Liebigs Ann.* **1995**, 855.

47. Bach, T.; Jödicke, K.; Kather, K.; Hecht, J. *Angew. Chem. Int. Ed. Engl.* **1995**, *34*, 2271.
48. Abe, M.; Ikeda, M.; Shirodai, Y.; Nojima, M. *Tetrahedron Lett.* **1996**, *37*, 5901.
49. Fukuzumi, S.; Fujita, M.; Otera, J. *J. Org. Chem.* **1993**, *58*, 5405.
50. Fleming, S. A.; Gao, J. J. *Tetrahedron Lett.* **1997**, *38*, 5407.
51. Desvergne, J.-P.; Bitit, N.; Castellan, A.; Webb, M.; Bouas-Laurent, H. *J. Chem. Soc. Perkin Trans. 2* **1988**, 1885.
52. Nakanishi, K.; Mizuno, K.; Otsuji, Y. *J. Chem. Soc. Perkin Trans. 1* **1990**, 3362.
53. Fleming, S. A.; Ward, S. C. *Tetrahedron Lett.* **1992**, *33*, 1013.
54. Bradford, C. L.; Fleming, S. A.; Ward, S. C. *Tetrahedron Lett.* **1995**, *36*, 4189.
55. Crimmins, M. T.; Guise, L. E. *Tetrahedron Lett.* **1994**, *35*, 1657.
56. Mizuno, K.; Hashizume, T.; Otsuji, Y. *J. Chem. Soc. Chem. Commun.* **1983**, 772.
57. Nakanishi, K.; Mizuno, K.; Otsuji, Y. *J. Chem. Soc. Chem. Commun.* **1991**, 90.
58. Kyushin, S.; Masuda, Y.; Matsushita, K.; Nakadaira, Y.; Ohashi, M. *Tetrahedron Lett.* **1990**, *31*, 6395.
59. Kyushin, S.; Nakadaira, Y.; Ohashi, M. *Chem. Lett.* **1990**, 2191.
60. Mizuno, K.; Ikeda, M.; Otsuji, Y. *Tetrahedron Lett.* **1985**, *26*, 461.
61. Kyushin, S.; Ehara, Y.; Nakadaira, Y.; Ohashi, M. *J. Chem. Soc. Chem. Commun.* **1989**, 279.
62. Mizuno, K.; Nishiyama, T.; Takahashi, N.; Inoue, H. *Tetrahedron Lett.* **1996**, *37*, 2975.
63. Mizuno, K.; Konishi, G.; Nishiyama, T.; Inoue, H. *Chem. Lett.* **1995**, 1077.
64. Konishi, G.; Chiyonobu, K.; Sugimoto, A.; Mizuno, K. *Tetrahedron Lett.* **1997**, *38*, 5313.
65. Mizuno, K.; Terasaka, K.; Ikeda, M.; Otsuji, Y. *Tetrahedron Lett.* **1985**, *26*, 5819.
66. Mizuno, K.; Nishiyama, T.; Terasaka, K.; Yasuda, M.; Shima, K.; Otsuji, Y. *Tetrahedron* **1992**, *48*, 9673.
67. Kubo, Y.; Todani, T.; Inoue, T.; Ando, H.; Fujiwara, T. *Bull. Chem. Soc. Jpn.* **1993**, *66*, 541.
68. Ohga, K.; Yoon, U. C.; Mariano, P. S. *J. Org. Chem.* **1984**, *49*, 213.
69. Mizuno, K.; Ikeda, M.; Otsuji, Y. *Chem. Lett.* **1988**, 1507.
70. Viehe, H. G. *Angew. Chem. Int. Ed. Engl.* **1979**, *18*, 917.
71. Viehe, H. G. *Acc. Chem. Res.* **1985**, *18*, 148.
72. Tamai, T.; Mizuno, K.; Hashida, I.; Otsuji, Y. *Chem. Lett.* **1992**, 781.
73. Tamai, T.; Mizuno, K.; Hashida, I.; Otsuji, Y. *Bull. Chem. Soc. Jpn.* **1993**, *66*, 3747.
74. Sulpizio, A.; Albini, A.; d'Alessandro, N.; Fasani, E.; Pietra, S. *J. Am. Chem. Soc.* **1989**, *111*, 5773.
75. d'Alessandro, N.; Fasani, E.; Mella, M.; Albini, A. *J. Chem. Soc. Perkin Trans. 2* **1991**, 1977.
76. d'Alessandro, N.; Albini, A.; Mariano, P. S. *J. Org. Chem.* **1993**, *58*, 937.

77. Fasani, E.; d'Alessandro, N.; Albini, A.; Mariano, P. S. *J. Org. Chem.* **1994**, *59*, 829.
78. Dinnocenzo, J. P.; Farid, S.; Goodman, J. L.; Gould, I. R.; Mattes, S. L.; Todd, W. P. *J. Am. Chem. Soc.* **1989**, *111*, 8973.
79. Todd, W. P.; Dinnocenzo, J. P.; Farid, S.; Goodman, J. L.; Gould, I. R. *Tetrahedron Lett.* **1993**, *34*, 2863.
80. Fukuzumi, S.; Okamoto, T.; Otera, J. *J. Am. Chem. Soc.* **1994**, *116*, 5503.
81. Abe, M.; Oku, A. *J. Chem. Soc. Chem. Commun.* **1994**, 1673.
82. Abe, M.; Oku, A. *J. Org. Chem.* **1995**, *60*, 3065.
83. Abe, M.; Nojima, M.; Oku, A. *Tetrahedron Lett.* **1996**, *37*, 1833.
84. Lin, X.; Kavash, R. W.; Mariano, P. S. *J. Org. Chem.* **1996**, *61*, 7335.
85. Khim, S.-K.; Cederstrom, E.; Ferri, D. C.; Mariano, P. S. *Tetrahedron* **1996**, *52*, 3195.
86. Hasegawa, E.; Brumfield, M. A.; Mariano, P. S.; Yoon, U. C. *J. Org. Chem.* **1988**, *53*, 5435.
87. Yoon, U. C.; Mariano, P. S. *Acc. Chem. Res.* **1992**, *25*, 233.
88. Kim, J.-M.; Hoegy, S. E.; Mariano, P. S. *J. Am. Chem. Soc.* **1995**, *117*, 100.
89. Yoon, U. C.; Kim, D. U.; Lee, C. W.; Choi, Y. S.; Lee, Y.-J.; Ammon, H. L.; Mariano, P. S. *J. Am. Chem. Soc.* **1995**, *117*, 2698.
90. Yoon, U. C.; Oh, S. W.; Lee, C. W. *Heterocycles* **1995**, *41*, 2665.
91. Hintz, S.; Fröhlich, R.; Mattay, J. *Tetrahedron Lett.* **1996**, *37*, 7349.
92. Heidbreder, A.; Mattay, J. *Tetrahedron Lett.* **1992**, *33*, 1973.
93. Mizuno, K.; Nakanishi, K.; Chosa, J.; Nguyen, T.; Otsuji, Y. *Tetrahedron Lett.* **1989**, *30*, 3689.
94. Mizuno, K.; Nakanishi, K.; Chosa, J.; Otsuji, Y. *J. Organomet. Chem.* **1994**, *473*, 35.
95. Kako, M.; Morita, T.; Torihara, T.; Nakadaira, Y. *J. Chem. Soc. Chem. Commun.* **1993**, 678.
96. Fukuzumi, S.; Kitano, T.; Mochida, K. *J. Chem. Soc. Chem. Commun.* **1990**, 1236.
97. Fukuzumi, S.; Kitano, T.; Mochida, K. *Chem. Lett.* **1989**, 2177.
98. Igarashi, M.; Ueda, T.; Wakasa, M.; Sakaguchi, Y. *J. Organomet. Chem.* **1991**, *421*, 9.
99. Nakadaira, Y.; Sekiguchi, A.; Funada, Y.; Sakurai, H. *Chem. Lett.* **1991**, 327.
100. Sakurai, H.; Sakamoto, K.; Kira, M. *Chem. Lett.* **1984**, 1213.
101. Watanabe, H.; Kato, M.; Tabei, E.; Kuwabara, H.; Hirai, N.; Sato, T.; Nagai, Y. *J. Chem. Soc. Chem. Commun.* **1986**, 1662.
102. Nakadaira, Y.; Komatsu, N.; Sakurai, H. *Chem. Lett.* **1985**, 1781.
103. Akasaka, T.; Ando, W.; Kobayashi, K.; Nagase, S. *J. Am. Chem. Soc.* **1993**, *115*, 10366.
104. Mikami, K.; Matsumoto, S.; Ishida, A.; Takamuku, S.; Suenobu, T.; Fukuzumi, S. *J. Am. Chem. Soc.* **1995**, *117*, 11134.
105. Kusukawa, T.; Shike, A.; Ando, W. *Tetrahedron* **1996**, *52*, 4995.
106. Kusukawa, T.; Ando, W. *Organometallics* **1997**, *16*, 4027.

107. Baciocchi, E.; Rol, C.; Rosato, G. C.; Sebastiani, G. V. *J. Chem. Soc. Chem. Commun.* **1992**, 59.
108. Mizuno, K.; Yasueda, M.; Otsuji, Y. *Chem. Lett.* **1988**, 229.
109. Mizuno, K.; Tamai, T.; Hashida, I.; Otsuji, Y. *J. Org. Chem.* **1995**, *60*, 2935.
110. Akasaka, T.; Sato, K.; Kako, M.; Ando, W. *Tetrahedron Lett.* **1991**, *32*, 6605.
111. Ando, W.; Kako, M.; Akasaka, T. *Chem. Lett.* **1993**, 1679.
112. Kato, N.; Takeda, K.; Nagasaki, Y.; Kata, M. *Ind. Eng. Chem. Res.* **1994**, *33*, 417.
113. Tamai, T.; Hashida, I.; Ichinose, N.; Kawanishi, S.; Inoue, H.; Mizuno, K. *Polymer* **1996**, *37*, 5525.
114. Ichinose, N.; Tamai, T.; Kawanishi, S.; Hashida, I.; Mizuno, K. *Langmuir* **1997**, *13*, 2603.
115. Mizuno, K.; Kobata, T.; Maeda, R.; Otsuji, Y. *Chem. Lett.* **1990**, 1821.
116. Mizuno, K.; Nakanishi, K.; Kobata, T.; Sawada, Y.; Otsuji, Y. *Chem. Lett.* **1993**, 1349.
117. Pandey, G.; Sesha, K. S.; Rao, P. *Angew. Chem. Int. Ed. Engl.* **1995**, *34*, 2669.
118. Kani, R.; Nakano, Y.; Majima, Y.; Hayase, S. *Macromolecules* **1996**, *29*, 4187.
119. Wang, Y.; West, R.; Yuan, C. *J. Am. Chem. Soc.* **1993**, *115*, 3844.
120. Caldwell, R. A.; Creed, D. *Acc. Chem. Res.* **1980**, *13*, 45.

INDEX

www.ingramcontent.com/pod-product-compliance
Ingram Content Group UK Ltd.
Pitfield, Milton Keynes, MK11 3LW, UK
UKHW020820290926
15470UKWH00003B/1275

9 780762 302130